Synthesis Lectures on Chemical Engineering and Biochemical Engineering

This series publishes short books on all aspects of chemical engineering, covering the analysis or design of chemical processes to effectively convert materials into more useful materials or energy. The books will focus on fundamental aspects necessary for chemical engineering design including chemistry, math, physics, and sometimes biology to improve the quality of life by inventing, optimizing, and economizing new technologies and products.

Atul H. Bari · Hemantkumar N. Akolkar ·
Navnath T. Hatvate · A. K. Haghi

Eco-Friendly Textile Processes

Springer

Atul H. Bari
Institute of Chemical Technology, Mumbai, Marathwada Campus
Jalna, Maharashtra, India

Navnath T. Hatvate
Institute of Chemical Technology, Mumbai, Marathwada Campus
Jalna, Maharashtra, India

Hemantkumar N. Akolkar
Department of Chemistry
Abasaheb Marathe College
Rajapur, Maharashtra, India

A. K. Haghi
Department of Chemistry
Institute of Molecular Sciences
University of Coimbra
Coimbra, Portugal

ISSN 2327-6738 ISSN 2327-6746 (electronic)
Synthesis Lectures on Chemical Engineering and Biochemical Engineering
ISBN 978-3-032-08915-1 ISBN 978-3-032-08916-8 (eBook)
https://doi.org/10.1007/978-3-032-08916-8

This Springer imprint is published by the registered company Springer Nature Switzerland AG
The registered company address is: Gewerbestrasse 11, 6330 Cham, Switzerland

Preface

The textile industry, one of the oldest and most dynamic sectors, has shaped economies, cultures, and societies for centuries. From its roots in natural fibers to today's global, technology-driven enterprise, textiles remain central to trade and livelihoods. However, this progress comes at a cost, as the industry is highly resource intensive, consuming vast amounts of water and energy while generating pollution and waste.

This book, *Eco-Friendly Textile Processing*, addresses these challenges and explores pathways for sustainable transformation. It highlights the industry's environmental footprint and presents advances in energy and water efficient technologies, green chemistry solutions, and innovative smart fabrics. Together, the chapters offer a comprehensive view of how textiles can evolve from a resource heavy model to one that is cleaner, circular, and resilient.

Chapter 1 sets the context by introducing the global significance of the textile industry, its complex value chain, and the pressing sustainability challenges such as water pollution, energy demand, air emissions, and waste generation. It also highlights pathways such as Green Supply Chain Management, certifications, and Corporate Social Responsibility initiatives that frame the broader sustainability discourse.

Chapter 2 focuses on drying technologies, one of the most energy intensive operations in textile processing. It explores eco-friendly alternatives including heat recovery systems, renewable heating sources, infrared drying, microwave-assisted techniques, and hybrid systems that demonstrate significant potential for reducing energy consumption and emissions.

Chapter 3 examines eco-friendly chemical treatment methods, particularly in wet processing. It highlights how green chemistry, enzyme-based processes, supercritical CO_2 dyeing, plasma treatment, and bio-based finishing agents can reduce chemical loads and effluent toxicity while maintaining or improving fabric quality and performance.

Chapter 4 looks ahead to emerging trends, showcasing the potential of smart textiles and advanced materials that combine functionality with sustainability. From biopolymers and nanomaterials to responsive fabrics and digital integration, this chapter envisions a future where innovation and environmental responsibility go hand in hand.

Together, these chapters provide a structured exploration of sustainable textile processing. It is hoped that this book will serve as a resource for researchers, industry practitioners, policymakers, and students committed to fostering a textile sector that not only sustains economies but also safeguards the environment for future generations.

Jalna, India	Dr. Atul H. Bari
Rajapur, India	Dr. Hemantkumar N. Akolkar
Jalna, India	Dr. Navnath T. Hatvate
Coimbra, Portugal	Dr. A. K. Haghi

Contents

About the Authors

Dr. Atul H. Bari, Ph.D. is an assistant professor at the Institute of Chemical Technology, Mumbai, Marathwada Campus, Jalna, Maharashtra, India. He holds a masters and a Ph.D. in of chemical engineering from the Department of Chemical Engineering at the Institute of Chemical Technology, Mumbai, Maharashtra, India. He has 15 years of research and 5 years of teaching experience. His areas of expertise and research interests include the mathematical modelling and simulations, population balance modelling for particulate processes, wastewater treatment, process intensification, crystallization, nanoparticle synthesis using flow synthesis, chemical kinetics. Dr. Bari has published over 17 research publications in international journals.

Hemantkumar N. Akolkar, Ph.D. is an associate professor in the Department of Chemistry at Rayat Shikshan Sanstha's Abasaheb Marathe Arts and New Commerce Science College, Maharashtra, India. He is an academician and researcher who has taught for the last 15 years at both undergraduate and postgraduate levels. He has published over 65 research papers in national and international journals and conferences of repute. Also, he has published 6 Indian patents and 7 research-oriented books with international publishers like Springer, Royal Society of Chemistry, IGI Global, etc. His areas of interest are heterocyclic chemistry, synthetic organic chemistry, and green chemistry. He is a reviewer of several journals of international repute in chemistry.

Dr. Navnath T. Hatvate, Ph.D. is an assistant professor at the Institute of Chemical Technology, Mumbai, Marathwada Campus, Jalna. He holds a Master of Pharmacy and a Ph.D. from the Department of Pharmaceutical Sciences and Technology at the Institute of Chemical Technology, Mumbai. He has 12 years of research and 6 years of teaching experience. His areas of expertise and research interests include the development of new chemical entities (NCE) and their analogues, modifications of excipients and their applications in drug delivery, the use of AI and ML in drug discovery and drug delivery, process chemistry of APIs and intermediates, and total synthesis of natural products, bioactive compounds, application of novel technologies in organic synthesis. Dr. Hatvate has published over 50 research publications and authored 4 books in international journals. He serves as an editorial board member for *Scientific Reports* and *Discover Pharmaceutical Sciences* (Nature Publishing).

A. K. Haghi, Ph.D. is a retired professor and has written, co-written, edited, or co-edited more than 1000 publications, including books, book chapters, and papers in refereed journals with over 4500 citations and H-index of 35, according to the Google Scholar database. Professor Haghi holds a B.Sc. in urban and environmental engineering from the University of North Carolina (USA) and has two M.Sc. degrees, one in mechanical engineering from North Carolina State University (USA) and another in applied mechanics, acoustics, and materials from the Université de Technologie de Compiègne (France). He was awarded a Ph.D. in engineering sciences at Université de Franche-Comté (France). He is a regular reviewer of leading international journals. He is appointed Honorary Research Associate (HRA) at the University of Coimbra, Portugal.

Textile Industry and Sustainability Context 1

Contents

Abstract

The textile industry is a key driver of trade and growth but is also highly resource-intensive and environmentally impactful. This chapter introduces the structure of the textile industry and outlines its key processing stages, beginning with raw materials, followed by dry processing and wet processing steps that transform fibers into finished fabrics. Specific emphasis is placed on the environmental footprint associated with each stage of production. The chapter tries to highlights the major sustainability challenges like water pollution from textile effluents, heat and energy requirements, air pollution and greenhouse gas emissions from energy-intensive operations and the generation of solid waste and micro-plastics during fiber processing. To address these issues, pathways to sustainability in textiles needs to be explored. By positioning environmental responsibility at the core of textile manufacturing, the industry can transition towards

A. H. Bari et al., *Eco-Friendly Textile Processes*, Synthesis Lectures on Chemical Engineering and Biochemical Engineering, https://doi.org/10.1007/978-3-032-08916-8_1

more sustainable practices while maintaining its global significance. This chapter provides a comprehensive overview of the textile processing chain and its associated sustainability context, setting the foundation for deeper exploration of solutions in subsequent chapters.

Keywords

Textile industry • Eco-friendly processing • Sustainability • Water pollution • Environmental impact

1.1 Introduction to the Textile Industry

The textile industry is one of the oldest and most active sectors of human economic activity. It's roots stretch back to prehistoric times when fibers from plants and animals were used for making fabrics for clothing and shelter. Clothing is no longer just for basic use or appearance; it now protects the body, reflects identity, profession, wealth, and power, and is one of the most important products traded worldwide. With the advent of the Industrial Revolution in the eighteenth century, textile manufacturing transformed into a mechanized and large-scale enterprise. The invention of spinning machinery, such as the spinning jenny and water frame, and weaving innovations like the power loom, accelerated productivity and contributed significantly to the rise of textile industries [3]. The textile and clothing industry has often been the first step on the industrialization ladder for many countries, enabling rapid growth in output [2]. It led to significant growth of countries like England, the USA, and Japan through industrialization and worldwide trade of the textile products. Even now, textile industry plays a crucial role in the growth of the developing economies like India and China [9, 17]. Countries such as Bangladesh, Sri Lanka, Vietnam, and Mauritius have also leveraged textile industry to move toward middle-income status [7]. This growth is largely driven by the industry's ability to absorb large numbers of unskilled workers while requiring relatively low start-up investment [21].

Today, the global textile market is geographically concentrated, with a small group of countries accounting for the majority of production and exports. China has emerged as the dominant producer, accounting for more than 47% of global textile and apparel share in 2005 [20], which was supported by economies of scale, infrastructure, and integrated supply chains. India follows as a leading hub, particularly in cotton textiles and home furnishings, with the sector playing a vital role in national employment and export earnings.

In the modern era, the textile sector has expanded into a globalized network characterized by complex value chains spanning raw material production, spinning, weaving, dyeing, finishing, and garment manufacturing. Cotton, polyester, and blends of natural and synthetic fibers dominate global consumption patterns. According to recent market analyses, the textile and apparel industry accounts for over USD 1.5 trillion in annual trade, and is expected to reach above USD 2.25 trillion by 2025 [13, 24] making it one of the largest segments of global manufacturing. The increasing diversification of fibers, including technical textiles and sustainable alternatives, continues to reshape market dynamics.

Sustainability challenges are increasingly shaping the trajectory of the industry. The sector is resource-intensive, consuming substantial amounts of water and energy in dyeing, finishing, and wet processing, and contributing significantly to greenhouse gas emissions [14]. The textile industry is estimated to employ over 60 million people and annually consume about 1 trillion kWh of electricity and 9 trillion liters of water to produce 60 billion kilograms of fabrics [25]. With rising consumer awareness and regulatory pressures, global brands and producers are investing in eco-friendly processing technologies, circular economy models, and sustainable supply chain initiatives [28]. Green Supply Chain Management (GSCM) integrates environmental considerations into all stages of the supply chain, from product design and material sourcing to manufacturing, logistics, and recycling [34]. In the textile industry, GSCM helps protect the environment while improving a company's brand image. To implement it effectively, firms often follow global certification systems such as ISO 14001:2015, Blue Sign®, and OEKO-TEX® Standard 100. Corporate Social Responsibility (CSR) meaning conducting business without harming society or the environment [30], is another initiative of companies for sustainability. In textiles, it encourages firms to adopt eco-friendly practices and investments. These transformations highlight the textile industry's dual role as both a driver of global economic growth and a focal point of debates on sustainable development.

In summary, the textile industry reflects a rich historical evolution and remains a cornerstone of the global economy. Its significance extends beyond trade and employment, influencing cultural exchange, technological innovation, and sustainability efforts worldwide.

1.2 Textile Processing Chain

The textile industry is a large and complex sector that converts raw fibers into finished fabrics and garments ready for consumer use. This transformation does not occur in a single step, but rather through a structured series of processes that gradually modify the properties of fibers to improve their usability, appearance, and performance. The processing chain broadly consists of raw material preparation, dry processing steps such as spinning and weaving, and wet processing steps such as dyeing, printing, and finishing [33]. Each stage has its own importance and together they form the backbone of the textile manufacturing cycle. Understanding this chain is essential to appreciate both the technical and environmental aspects of the textile industry (Fig. 1.1).

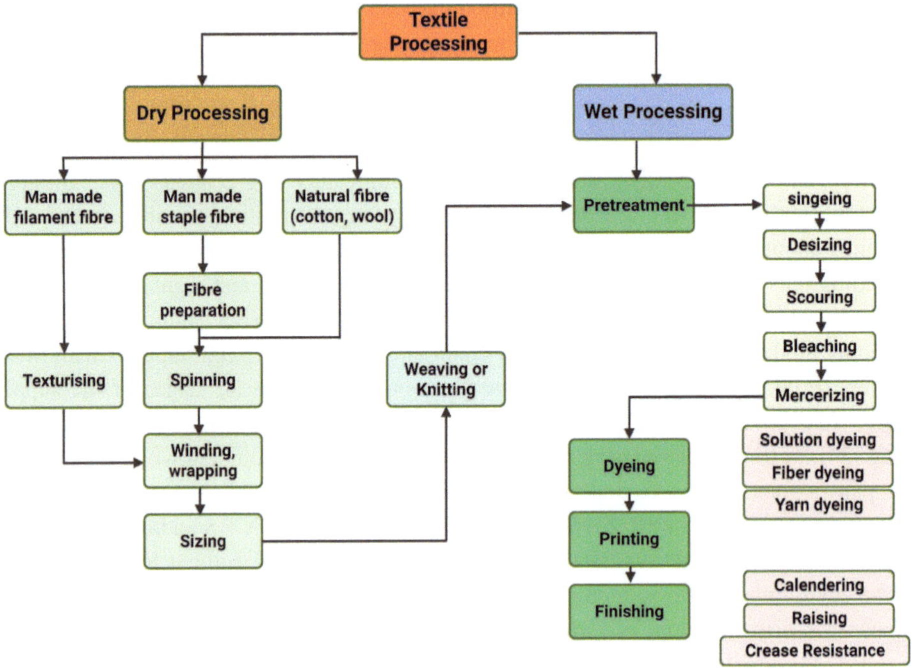

Fig. 1.1 Typical textile processing chain

1.2.1 Raw Materials

The textile journey begins with raw materials. Textiles can be made from natural fibers such as cotton, wool, silk, and flax, or man-made fibers like polyester, nylon, acrylic, and viscose. Natural fibers are derived from plants or animals, while man-made fibers are produced through chemical and industrial processes. Cotton remains the most widely used natural fiber because of its softness, breathability, and comfort. It accounts for over 39% of the raw material used in the textile industry [32]. Wool and silk, though used in smaller quantities, are prized for warmth and luxury. On the other hand, synthetic fibers such as polyester and nylon dominate global textile production due to their durability, low cost, and versatility [4].

Within man-made fibers, two broad categories exist: filament fibers and staple fibers. Filament fibers are long continuous fibers, like polyester filament yarn, that can be directly woven or knitted. Staple fibers, in contrast, are short in length and need to undergo processes like spinning to convert them into yarn. Natural fibers like cotton and wool naturally occur as staple fibers, while synthetic fibers can be engineered into either filament or staple form. The selection of fiber type at this stage has a major influence on the later steps of processing and the final fabric quality.

1.2.2 Dry Processing

After the raw fibers are obtained, they enter the dry processing stage, which focuses on converting fibers into yarns and then into fabrics [33]. These stages may be collectively referred as "dry" because it does not involve water or chemical treatments, but rather mechanical and physical processes. For man-made filament fibers, the main dry processing step is texturizing [11]. Filament yarns are often too smooth and uniform, which can make fabrics feel flat and uncomfortable. Texturizing introduces crimps, coils, or loops into the filament, improving its bulk, elasticity, and hand feel. Texturized yarns are commonly used in sportswear, hosiery, and stretch fabrics. For man-made staple fibers and natural fibers like cotton and wool, the process begins with fiber preparation. Here, fibers are cleaned, opened, and blended to remove impurities and create a uniform fiber mix. This is followed by spinning, where fibers are twisted together to form yarns of the desired thickness and strength. Modern spinning methods include ring spinning, rotor spinning, and air-jet spinning, each offering different yarn qualities and production speeds. Spinning is highly energy intensive and may consumes up to 34% energy from entire textile industry [8]. Spinning also generates solid waste which can be divided in two main types: dust and solid particles removed during carding and combing. Since yarn realization is usually only 75–80%, about 20–25 kg of waste is produced for every 100 kg of material processed, including some invisible losses [8].

Once yarns are spun, they often undergo winding and wrapping, which prepares them for weaving or knitting. Yarns are wound onto bobbins, cones, or beams in a form that can be easily handled by fabric-forming machines. For woven fabrics, sizing is a critical step. In sizing, yarns are coated with a thin film of starch or synthetic polymers to strengthen them and reduce breakage during weaving. Without sizing, the yarns would be too weak to withstand the tension and friction in the loom. The yarns then move into the fabric formation stage, where they are converted into textiles through either weaving or knitting. Weaving interlaces warp and weft yarns at right angles on a loom, producing strong, stable fabrics such as denim, poplin, and canvas. Knitting, on the other hand, uses interloping yarns to create fabrics with more elasticity and comfort, such as t-shirts, sweaters, and hosiery. At the end of dry processing, raw grey fabrics are produced. These fabrics have the basic structure but lack the desired color, feel, and functional properties. To achieve these qualities, they need to undergo wet processing.

1.2.3 Wet Processing

Wet processing is the stage where fabrics are treated with water, chemicals, and heat to improve their aesthetics, durability, and functionality. It transforms raw grey fabrics into finished textiles with the desired look and performance. This stage can be broadly divided into pretreatment, dyeing, printing, and finishing.

1.2.3.1 Pretreatment is the first step of wet processing. It removes impurities and prepares the fabric for uniform dyeing and finishing. Pretreatment includes several operations:

(i) Singeing: Singeing is the process of burning off protruding fibers from the surface of a fabric to achieve a smooth and clean appearance. It is usually done by passing the fabric over a gas flame or heated copper plates. This process improves fabric quality by reducing pilling, enhancing luster, and providing a uniform surface for subsequent treatments like dyeing, printing, and finishing. Singeing is essential for cotton and blends.

(ii) Desizing: Desizing removes the size materials (starch, gums, synthetic agents) applied to warp yarns during weaving to improve strength and reduce breakage. Since these materials hinder wet processing, they must be eliminated before scouring and bleaching. Desizing is carried out using enzymatic, oxidative, or acid methods, depending on the type of size applied. The process improves absorbency, facilitates uniform dyeing, and prepares the fabric for further wet processing steps. However, typical desizing would result in load of 45.5 kg biochemical oxygen demand (BOD) and 91 kg chemical oxygen demand (COD) for every 1000 kg of cotton fabrics [23].

(iii) Scouring: Scouring is an alkaline treatment designed to remove natural impurities such as waxes, oils, pectins, and residual dirt from textile fibers. It enhances the fabric's wettability and absorbency, making it receptive to dyes and finishes. Typically, scouring is done using caustic soda under controlled conditions. Proper scouring ensures uniform dye uptake, prevents patchy coloring, and contributes to the overall brightness and cleanliness of the fabric. Scouring typically consumes 20–45 m^3 of water per Ton of product [15].

(iv) Bleaching: Bleaching is the chemical process of whitening textiles by destroying natural coloring matter present in fibers. Common bleaching agents include hydrogen peroxide, sodium hypochlorite, and sodium chlorite. This step enhances brightness and provides a neutral base shade for dyeing or printing. Controlled bleaching ensures that the strength of the fibers is not compromised while achieving the desired degree of whiteness. It is a critical step in preparing textiles for light or pastel shades. Bleaching typically consumes 2.5–25 m^3 of water per Ton of product [15].

(v) Mercerizing: Mercerizing is a treatment applied mainly to cotton fabrics using concentrated sodium hydroxide (caustic soda) under tension. This process swells the fibers, increasing luster, strength, smoothness, and dye affinity. The fabric develops a silk-like appearance and improved dimensional stability. Mercerized cotton absorbs dyes more evenly and deeply, resulting in brighter, more vibrant shades. It is a valuable finishing process that significantly enhances both the aesthetic and functional qualities of cotton textiles.

1.2.3.2 Dyeing: Dyeing is the process of imparting color to textiles by applying dyes or pigments that bond with fibers. It can be carried out at different stages of textile manufacture. Solution dyeing involves adding pigments to the polymer solution before fiber extrusion, ensuring excellent colorfastness and uniformity. Fiber dyeing (stock dyeing) colors fibers before spinning, producing mélange effects and heather tones. Yarn dyeing is done after spinning but before weaving or knitting, allowing creation of stripes, checks, or jacquards. Each method serves specific design or performance needs, with solution dyeing often used for synthetics, and yarn dyeing commonly applied in shirting, suiting, and decorative fabrics. Proper dyeing enhances textile aesthetics, durability, and consumer appeal while maintaining uniformity and fastness to washing, rubbing, and light. Textile dyeing has been linked to around 72 toxic chemicals in water, including 30 that are difficult to remove [16]. As per the World Bank estimation, about 17–20% of industrial waste water generated is from textile dyeing and finishing treatment [12].

1.2.3.3 Printing: Printing is a localized coloration process in which dyes or pigments are applied in specific patterns or motifs on fabric surfaces. Unlike dyeing, which colors the entire substrate, printing creates decorative effects, often with multiple colors. Traditional methods include block printing, roller printing, and screen printing, while modern approaches such as digital printing offer high precision and limitless design flexibility. Pigment printing is widely used for its ability to create sharp, durable patterns without requiring post-dye fixation. Reactive and disperse dye printing produce vibrant, wash-fast colors on cotton and synthetics, respectively. Printing enhances textile value by combining functionality with fashion, enabling intricate designs, gradients, and multicolor effects. Applications range from apparel and home textiles to technical fabrics. High-quality printing depends on fabric preparation, proper fixation, and finishing to ensure durability and aesthetic appeal.

1.2.3.4 Finishing: Finishing refers to the range of chemical and mechanical treatments applied after dyeing or printing to improve fabric appearance, performance, and handle. Calendering passes fabric through heated rollers, producing a smooth, glossy surface or embossed designs, enhancing luster and drape. Raising (or napping) lifts surface fibers using teasels or wire rollers, creating a soft, warm texture as in flannels and blankets. Crease resistance finishing involves applying resins or crosslinking agents to reduce wrinkling and improve dimensional stability, particularly in cotton fabrics. Finishing also enhances properties like water repellency, flame retardancy, or softness, depending on end use. These treatments add value and functionality, making textiles more attractive, comfortable, and durable. Effective finishing balances aesthetics, performance, and sustainability by minimizing chemical use and maintaining fabric strength.

1.2.3.5 Drying: Drying is a secondary but crucial step in wet processing, occurring repeatedly after pretreatment, dyeing, printing, and certain finishing processes.

In solution dyeing, fiber dyeing, or yarn dyeing, drying is essential to fix the dye and prepare the material for subsequent steps like winding, weaving, or printing. Once printing pastes are applied, the fabric is usually dried before fixation through steaming or curing. Some finishing operations, such as creasing resistance, raising, or resin finishes, involve applications of wet chemical. Drying is also necessary here to stabilize the finish. Drying is highly energy intensive and it may consumes up to 80% of energy consumed in textile industry [5].

1.3 Environmental and Sustainability Challenges

The textile industry is one of the most resource-intensive and polluting sectors, exerting pressure on water, energy, and ecosystems throughout its value chain. Every stage, from fiber cultivation and yarn manufacturing to dyeing, finishing, and garment care, contributes to an ecological footprint. While cotton farming alone demands high irrigation water and pesticide use; synthetic fibers such as polyester rely heavily on fossil fuels and contribute to long-term plastic pollution. In processing, particularly dyeing and finishing, large quantities of chemicals and heated water are used, leading to wastewater effluents with high toxicity, pH, salinity, and color load [19, 27, 29].

Air emissions add another dimension, with boilers, dryers, and finishing machines releasing greenhouse gases and particulate matter. Additionally, fugitive emissions of volatile organic compounds (VOCs) from printing and solvent-based coatings worsen indoor and outdoor air quality. Waste generation in the form of sludge, rejected fabric, and cutting waste adds to landfill burdens, while microplastics from synthetic fibers increasingly appear in aquatic environments [1].

These challenges underscore the industry's dual burden: on one hand, it provides livelihoods to millions; on the other, it generates critical environmental and sustainability problems. The need for systemic transition towards cleaner production, resource efficiency, and circularity have thus become critical. To implement sustainability practices effectively, it is important to understand the key environmental problems posed by the textile industry, which are outline briefly in the following subsections.

1.3.1 Water Pollution from Textile Effluents

Wet processing operations such as bleaching, dyeing, printing, and finishing are the dominant contributors to water pollution. A typical textile plant may discharge 100–150 L of water per kilogram of fabric processed [22]. This water when discharged is often laden with dyes, salts, surfactants, and heavy metals. Figure 1.2 shows the typical amount of

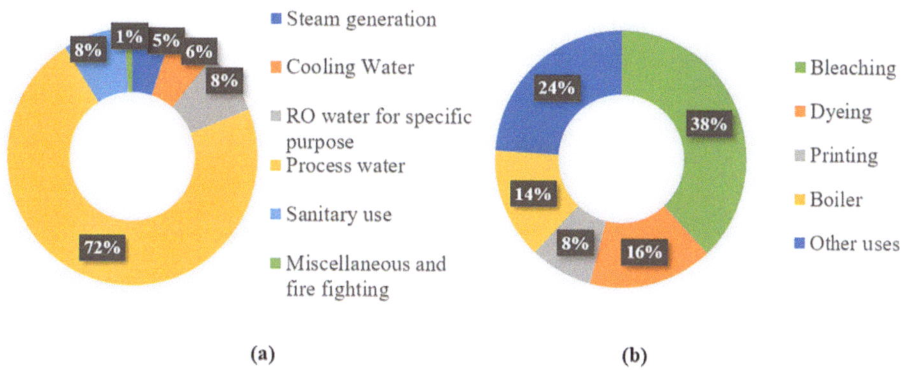

Fig. 1.2 Percentage water consumption **a** across various facilities and **b** breakdown in wet processing of textile industry

water required in various textile manufacturing process [10]. The effluents are characterized by high chemical oxygen demand (COD), biological oxygen demand (BOD), and persistent color, making them resistant to conventional biological treatment [19, 27, 29].

Synthetic dyes, especially azo dyes, pose additional hazards due to their breakdown into carcinogenic amines under anaerobic conditions. High temperature and pH discharges can also disrupt aquatic ecosystems, affecting fish and microbial diversity. Moreover, the widespread use of reactive dyes leads to poor fixation on cotton, causing substantial dye loss into wastewater streams.

While effluent treatment plants (ETPs) are mandated in many countries, their efficiency depends on proper operation and advanced technologies such as membrane filtration, ozonation, and advanced oxidation processes. Yet, small-scale textile clusters often lack financial and technical capacity to operate ETPs effectively. Hence, water pollution remains one of the gravest environmental challenges of textile processing, directly linked to sustainability concerns and regulatory scrutiny.

1.3.2 Heat and Energy Requirements

Textile processing is highly energy-intensive, particularly during wet-finishing stages like dyeing, drying, and heat-setting. In the textile industry, about 40–45% of total energy is used for yarn and fabric manufacturing, while 35–60% is consumed in wet processing [6]. In the textile industry, energy use is divided mainly into electricity and thermal energy. Electricity powers machinery, lighting, air conditioning, motor drives, and compressed air systems, and is the dominant source in spinning, weaving, and knitting. Thermal energy, mostly from steam generated by oil, LPG, coal, or natural gas, is mainly used in wet processing operations such as bleaching, scouring, dyeing, washing, and finishing, typically at

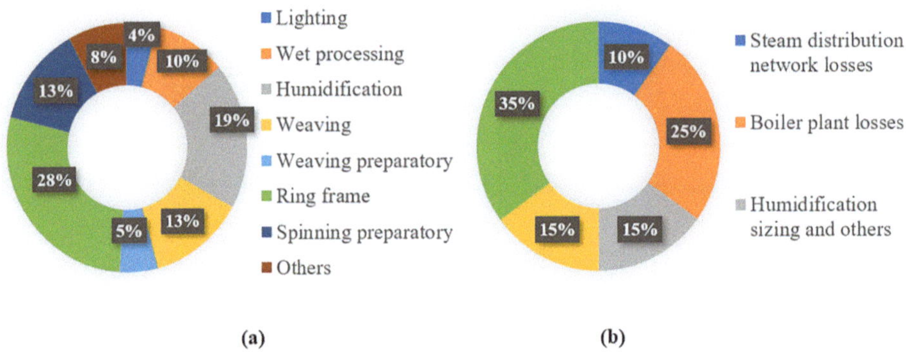

Fig. 1.3 Distribution of **a** power and **b** heat requirement in various processes of textile industry

temperatures of 60–200 °C. Boilers, stenters, and dryers are major thermal energy users. Stenter machines which is central to size control, finishing, and heat-setting often disperse significant heat through exhaust flue gases. The wet processing sector is the largest consumer of both energy and water, with water being a major cost factor. In garment manufacturing, electricity is used for cutting, sewing, inspection, and transport, while ironing requires thermal energy. Figure 1.3 shows typical power and heat distribution in various processes of textile industry [26].

1.3.3 Air Pollution and Greenhouse Gas Emissions

Air pollution in textiles arises from both direct fuel combustion and process related emissions. Boilers, stenters, dryers consume fossil fuels such as coal, oil, or natural gas, releasing CO_2, SO_2, NO_X, particulate matter. In developing economies where coal remains common, greenhouse gas emissions are particularly high. Printing operations using solvent-based inks coatings emit VOCs, contributing to smog formation occupational hazards. The carbon footprint of textile products is significant when considering their life cycle. For example, the production of one wear jeans may require up to 2.5 kg of CO_2 equivalent [18]. Dryers stenter machines are among the most energy-intensive operations and inefficient heat recovery systems magnify emissions.

Climate change concerns have brought increasing attention to the textile sector's greenhouse gases contribution, which is estimated at 6–8% of global emissions [14]. Sustainable energy management, fuel substitution (e.g., biomass or renewable electricity), adoption of cleaner process technologies are critical pathways to mitigate air pollution greenhouse gases emissions.

1.3.4 Solid Waste and Microplastic Generation

Solid waste streams from the textile industry include sludge from effluent treatment, fiber fabric waste from spinning garmenting, packaging residues. Sludge from dyeing finishing often contains high levels of heavy metals, dyes, salts, making its disposal problematic. In many cases, it is either landfilled or incinerated, both of which carry environmental burdens. The rise of synthetic fibers has added another dimension to microplastic release. During washing abrasion of polyester, nylon, acrylic fabrics, microfibers are shed into wastewater, eventually reaching rivers oceans. Studies show that textiles are one of the largest contributors to microplastic pollution globally, affecting marine life and potentially human health as well. Cutting waste from garment manufacturing, estimated up-to 19% of fabric input [31], represents not only a solid waste issue but also embedded resource loss. Recycling initiatives such as fiber-to-fiber recovery, mechanical re-spinning, chemical depolymerization are being explored, but their industrial uptake is still limited.

1.4 Pathways to Sustainability in Textiles

In textile industry, manufacturing point of view, identifying environmental concerns related to particular process is first step to attain sustainability. Addressing these environmental challenges will require integrated pathways that combine cleaner production, resource efficiency, systemic change. The key environmental concerns in textile processing and their possible sustainable approaches are summarized in Table 1.1. Overall, future sustainability in textiles will increasingly depend on combining eco-friendly smart materials, innovative fabrication techniques, and green chemistry approaches. Developments such as bio-based fibers, nanomaterials, and non-toxic functionalization, along with water and energy saving processes, support reduced environmental impact. Sustainable chemical processing methods and improved heat utilization and optimization are also essential for lowering emissions and resource use. At the same time, digitalization and Industry 4.0 tools like IoT, AI, and digital twins enable efficiency, lifecycle management, and circular economy integration.

Circular economy strategies like closed-loop recycling, extended producer responsibility (EPR), design for recyclability should be promoted by manufacturer. Certifications such as OEKO-TEX, GOTS, BCI provide assurance push the industry toward more responsible sourcing [22]. These approaches help to minimize waste and extend material life. On the consumer side, promoting sustainable laundry practices garment longevity campaigns also reduces lifecycle impacts. Finally, policy interventions, stakeholder collaboration, consumer awareness are vital enablers. Government regulations on effluent discharge, incentives for renewable energy adoption, industry-academia partnerships for innovation play key roles. With collective effort, the textile sector can transform from a resource-intensive model to a sustainable, low-impact industry.

Table 1.1 Environmental concerns sustainable approaches in textile processing

Textile process	Key resources used	Environmental concerns	Possible sustainable approaches
Fiber preparation for natural fibers	Water, energy for cleaning, detergents, alkali	High water use, wastewater with organic load, residue of pesticide (cotton)	Organic farming, efficient cleaning methods, water recycling
Texturizing of filament fibers	Electricity, compressed air, lubricants	High energy use, carbon footprint	Energy-efficient heaters, use of renewable energy
Spinning	Electricity for machines, lubricants, humidification water	High energy demand, oil waste, dust in air	Use of energy-efficient motors, waste recycling, dust control systems
Sizing	Starch, PVA, synthetic binders, water, heat	Wastewater with high COD BOD, polymer residues	Bio-based sizing agents, recycling of sizing materials, enzymatic desizing
Weaving	Electricity, sizing materials (for weaving)	Noise pollution, energy use, broken yarn particle waste	Automation, energy-efficient looms, recycling of yarn waste
Pretreatment processes (singeing, desizing, scouring, bleaching, mercerizing etc)	Water, caustic soda, acids, surfactants, oxidizing agents, energy	High water use, chemical-laden effluents, greenhouse gas emissions	Low-liquor ratio machines, enzymatic processes, closed-loop water systems
Dyeing	Large volumes of water, dyes, salts, energy for heating	Colored effluents, toxic chemicals, high salt load	Low-liquor dyeing, natural dyes, salt-free dyeing
Printing	Water, chemicals in the form of dyes pigments, thickeners	Effluent with dyes, pigments, use of high energy	Digital printing, improved fixation techniques, wastewater treatment
Finishing (calendering, raising, crease resistance, functional finishes)	Energy (for rollers, steam), resins, formaldehyde-based chemicals	VOCs emissions, formaldehyde release, high energy use	Eco-friendly resins, plasma finishing, use of renewable energy
Garment making (cutting and sewing)	Electricity, packaging materials	Fabric waste, solid waste from cut pieces, plastic packaging	Waste minimization, recycling, eco-friendly packaging

1.5 Concluding Remarks

The textile industry plays an important role in global trade, economy provides livelihoods to millions of people. At the same time, it creates serious environmental concerns due to its high demand for water, energy, chemicals. Pollution from wastewater, release of greenhouse gases, generation of solid waste are some of the most critical challenges. These impacts make it necessary for the textile industry to shift toward more sustainable and eco-friendly processing. By adopting cleaner technologies, improving efficiency in the use of resources, moving towards circular practices such as recycling reuse, the sector can greatly reduce its footprint.

This book highlights the necessary strategy of eco-friendly textile processing to balance economic growth with environmental care. In the following chapters, specific topics are discussed in more detail. These include heat processes in textile production, chemical treatment processes, new emerging trends in textile industry that point toward a more sustainable future.

References

1. Acharya, S., Rumi, S. S., Hu, Y., & Abidi, N. (2021). Microfibers from synthetic textiles as a major source of microplastics in the environment: A review. *Textile Research Journal, 91*(17–18), 2136–2156. https://doi.org/10.1177/0040517521991244
2. Brenton, P., & Hoppe, M. (2007). *Clothing and export diversification: Still Aroute to growth for low-income countries?* (Vol. 4343). World Bank Publications.
3. Chapman, S. D. (1972). The cotton industry in the industrial revolution. In *The Industrial Revolution A Compendium* (pp. 1–64). Springer.
4. Deopura, B., & Padaki, N. (2015). Synthetic textile fibres: Polyamide, polyester and aramid fibres. In *Textiles and fashion* (pp. 97–114). Elsevier.
5. Efremov, G. (2000). Generalized kinetics of drying of fibre materials. *Fibre Chemistry, 32*(6), 431–436.
6. Farhana, K., Kadirgama, K., Mahamude, A. S. F., & Mica, M. T. (2022). Energy consumption, environmental impact, and implementation of renewable energy resources in global textile industries: An overview towards circularity and sustainability. *Materials Circular Economy, 4*(1), 15. https://doi.org/10.1007/s42824-022-00059-1
7. Gereffi, G., Frederick, S., & Gereffi, G. (2010). *The global apparel value chain, trade and the crisis: Challenges and opportunities for developing countries.*
8. Goyal, A., & Nayak, R. (2020). Sustainability in yarn manufacturing. In *Sustainable Technologies for Fashion and Textiles* (pp. 33–55). Elsevier. https://doi.org/10.1016/B978-0-08-102867-4.00002-5
9. Haque, S. I., & Thaku, I. A. (2015). Development of textile industry of India and China: Flying geese model revisited. *Pacific Business Review International, 8*(2).
10. Hauser, P. (2011). *Textile Dyeing.* IntechOpen. https://books.google.co.in/books?id=3OqPDwAAQBAJ
11. Hearle, J. W., Hollick, L., & Wilson, D. K. (2001). *Yarn texturing technology.*

12. Holkar, C. R., Jadhav, A. J., Pinjari, D. V., Mahamuni, N. M., & Pandit, A. B. (2016). A critical review on textile wastewater treatments: Possible approaches. *Journal of Environmental Management, 182*, 351–366. https://doi.org/10.1016/j.jenvman.2016.07.090
13. Ikram, M. (2022). Transition toward green economy: Technological Innovation's role in the fashion industry. *Current Opinion in Green and Sustainable Chemistry, 37*, 100657. https://doi.org/10.1016/j.cogsc.2022.100657
14. Imran, S., Mujtaba, M., Zafar, M. M., Hussain, A., Mehmood, A., Farwa, U. E., Korakianitis, T., Kalam, M., Fayaz, H., & Saleel, C. A. (2023). Assessing the potential of GHG emissions for the textile sector: A baseline study. *Heliyon, 9*(11).
15. Ismail, G. A., & Sakai, H. (2022). Review on effect of different type of dyes on advanced oxidation processes (AOPs) for textile color removal. *Chemosphere, 291*, 132906. https://doi.org/10.1016/j.chemosphere.2021.132906
16. Kant, R. (2012). Textile dyeing industry an environmental hazard. *Natural Science, 04*(01), 22–26. https://doi.org/10.4236/ns.2012.41004
17. Kim, J.-O., Traore, M., & Warfield, C. (2006). The textile and apparel industry in developing countries. *Textile Progress, 38*(3), 1–64.
18. Li, Z., Zhou, Y., Zhao, M., Guan, D., & Yang, Z. (2024). The carbon footprint of fast fashion consumption and mitigation strategies-a case study of jeans. *Science of The Total Environment, 924*, 171508. https://doi.org/10.1016/j.scitotenv.2024.171508
19. Mani, S., Chowdhary, P., & Bharagava, R. N. (2019). Textile Wastewater Dyes: Toxicity Profile and Treatment Approaches. In R. N. Bharagava & P. Chowdhary (Eds.), *Emerging and Eco-Friendly Approaches for Waste Management* (pp. 219–244). Springer Singapore. https://doi.org/10.1007/978-981-10-8669-4_11
20. McCann, J. (2011). China's textile and apparel industry and the global market: Five competitive forces. *SAM Advanced Management Journal (Society for Advancement of Management), 76*(1).
21. Nordås, H. K. (2004). *The global textile and clothing industry post the agreement on textiles and clothing* (Issue 5). WTO discussion paper.
22. Palamutcu, S. (2017). Sustainable Textile Technologies. In S. S. Muthu (Ed.), *Textiles and Clothing Sustainability* (pp. 1–22). Springer Singapore. https://doi.org/10.1007/978-981-10-2474-0_1
23. Panda, S. K. B. C., Sen, K., & Mukhopadhyay, S. (2021). Sustainable pretreatments in textile wet processing. *Journal of Cleaner Production, 329*, 129725. https://doi.org/10.1016/j.jclepro.2021.129725
24. Papamichael, I., Chatziparaskeva, G., Pedreño, J. N., Voukkali, I., Almendro Candel, M. B., & Zorpas, A. A. (2022). Building a new mind set in tomorrow fashion development through circular strategy models in the framework of waste management. *Current Opinion in Green and Sustainable Chemistry, 36*, 100638. https://doi.org/10.1016/j.cogsc.2022.100638
25. Raja, A. S. M., Arputharaj, A., Saxena, S., & Patil, P. G. (2019). Water requirement and sustainability of textile processing industries. In *Water in Textiles and Fashion* (pp. 155–173). Elsevier. https://doi.org/10.1016/B978-0-08-102633-5.00009-9
26. Sathaye, J., Price, L., & Fridley, D. (2005). Assessment of energy use and energy savings potential in selected industrial sectors in India. *Bulletin on Energy Efficiency, 6*(2), 25–28.
27. Savin, I.-I., & Butnaru, R. (2008). WASTEWATER CHARACTERISTICS IN TEXTILE FINISHING MILLS. *Environmental Engineering and Management Journal, 7*(6), 859–864. https://doi.org/10.30638/eemj.2008.113
28. Shen, B. (2014). Sustainable fashion supply chain: Lessons from H&M. *Sustainability, 6*(9), 6236–6249.

29. Singha, K., Pandit, P., Maity, S., & Sharma, S. R. (2021). Harmful environmental effects for textile chemical dyeing practice. In *Green Chemistry for Sustainable Textiles* (pp. 153–164). Elsevier. https://doi.org/10.1016/B978-0-323-85204-3.00005-1
30. Steiner, G. A., Steiner, J. F., & Steiner, G. A. (1994). Business, government, and society: A managerial perspective: Text and cases. *(No Title).*
31. Taifa, I. W., Twaha, I., & Mwakibambo, M. A. (2021). Critical Analysis of Material Consumption and Cost Reduction Techniques for the Apparel Cutting Processes. *Tanzania Journal of Science, 47*(5), 1689–1700. https://doi.org/10.4314/tjs.v47i5.17
32. Uddin, F. (2019). Introductory Chapter: Textile Manufacturing Processes. In F. Uddin (Ed.), *Textile Manufacturing Processes.* IntechOpen. https://doi.org/10.5772/intechopen.87968
33. Vigo, T. L. (2013). *Textile processing and properties: Preparation, dyeing, finishing and performance* (Vol. 11). Elsevier.
34. Zhu, Q., Sarkis, J., & Geng, Y. (2005). Green supply chain management in China: Pressures, practices and performance. *International Journal of Operations & Production Management, 25*(5), 449–468.

Sustainable Heat Management and Drying in Textile Processing

2

Contents

Abstract

The textile industry is one of the most resource-intensive sectors, with drying processes accounting for up to a quarter of total energy consumption in textile finishing. Traditional drying techniques, which include convection, conduction, and radiation, are effective in removing moisture but are highly reliant on fossil fuels, hence contributing enormously to greenhouse gas emissions into the environment. This chapter dwells on some eco-friendly solutions that achieve high energy efficiency, ethically lower operational costs, and sustainable manufacturing. Innovations in convection drying, such as waste heat recovery, IoT moisture sensors, and hybrid solar-thermal drying, show

A. H. Bari et al., *Eco-Friendly Textile Processes*, Synthesis Lectures on Chemical Engineering and Biochemical Engineering, https://doi.org/10.1007/978-3-032-08916-8_2

significant savings in fuel consumption and emissions. Conduction drying also progresses using effective steam transport, condensation recovery, and renewable energy to heat, like biomass or solar thermal energy. Radiation drying, including infrared, microwave-aided, and hybrid microwave-convection units, can provide on-zone drying of a wide range of products with energy savings up to 77% compared to convection. Taken together, they reflect a paradigm shift in the processing of textiles as they focus on sustainability, integration of renewable energy, and the intelligent control of such processes to produce greener and more efficient drying technology.

Keywords

Textile drying • Eco-friendly technologies • Sustainability • Energy efficient processing

2.1 Introduction

The textile sector has fundamentally changed in recent years. Previously, most textile manufacturing could be found in developed economies that aimed to supply the domestic market. Nevertheless, the emergence of international trade and the internationalization of production have changed the textile industry. Nowadays, the production of textiles is moving towards developing countries because the production costs are cheaper in such places than in modernized countries. This has enabled textile firms to stay in the international market. Along with that, there is also the question of new technologies being involved in developing the textile industry. The textile sector is among the world's highest energy and water consumers. The most energy-demanding process in textile finishing is drying. Following wet operations, including washing, bleaching, dyeing, and printing, textiles must be dried to deliver a stable and market-ready product [35]. The drying of the fabrics is a complex process of heat and mass transfer into the fabric, resulting in physical and chemical modification. The three modes of heat transfer include: convection, conduction, and radiation. Conventional drying techniques consume disproportionate energy, about 12–25%. They are consequently costly to operate and hazardous to the environment, owing to their reliance on fossil fuels, the emissions of which contribute to the greenhouse effect. As a result, there has been a strong requirement to establish green drying technologies that help make the textile industry more sustainable.

Eco-friendly drying focuses on conserving energy, being easy on the environment, and maximizing the process's efficiency to improve the quality of the textiles. Some standard drying techniques include convection, conduction, and radiation. Recent developments, including the incorporation of renewable energy sources, the recovery of waste heat, and the use of hybrid drying systems, are making efforts to shape the drying process into a more sustainable part of the textile production process [22].

This chapter focuses on the fundamental of drying considering heat and mass transfer. Also different types of drying processes used in textiles are discussed along with their modes of operation, their eco-friendly modifications, and their sustainable implication to production.

2.1.1 Fundamentals of Heat and Mass Transfer in Textile Porous Media

Heat and mass transfer in porous fabrics represent a coupled transport problem where thermal energy and moisture simultaneously interact within a heterogeneous fibrous assembly. A fabric consists of fibers, yarns, pores, and entrapped air, forming a multiscale porous medium with complex transport pathways. Understanding fundamentals of conduction, convection, radiation, and diffusion within this system is essential for modeling drying, finishing, and thermo-mechanical stabilization processes in textiles.

The three basic modes of heat transfer: conduction, convection, and radiation are all present in textiles.

2.1.1.1 Heat transfer by Conduction: Conduction occurs along fibers, across yarn contacts, and through entrapped air pockets. Fourier's law governs conductive transport:

$$q_x = -k_{eff} \frac{\partial T}{\partial x} \tag{2.1}$$

where q_x is the heat flux in the x direction, and k_{eff} is the effective thermal conductivity of the textile. For fabrics, k_{eff} is a bulk property depending on fiber conductivity (k_f), air conductivity (k_a), and porosity (ε):

$$k_{eff} = (1 - \varepsilon)k_f + \varepsilon k_a \tag{2.2}$$

This effective model is simplistic; in reality, tortuosity and contact resistance modify the heat pathways.

2.1.1.2 Convective heat transfer: Convection dominates at fabric surfaces, especially in dryers where hot air interacts with moisture laden fabrics. Newton's law of cooling defines convective transfer:

$$q = hA(T_s - T_\infty) \tag{2.3}$$

with h as the convective heat transfer coefficient, A surface area, T_s fabric surface temperature, and T_∞ bulk air temperature. In porous fabrics, convective exchange also occurs internally when air penetrates pores under forced flow.

2.1.1.3 Radiative heat transfer: Radiation is particularly important in infrared heating and high-temperature processes. Net radiative exchange between surfaces is expressed as:

$$q = \varepsilon_{rad} \sigma A \left(T_s{}^4 - T_\infty{}^4 \right) \tag{2.4}$$

where ε_{rad} is fabric emissivity, σ is the Stefan–Boltzmann constant, and T_∞ is the surrounding enclosure temperature.

2.1.1.4 Mass transfer in porous fabrics:

Moisture transport within fabrics involves simultaneous liquid movement (capillary flow) and vapor diffusion.

(i) Diffusion of vapor follows Fick's law:

$$J = -D_{eff} \left(\frac{\partial C}{\partial x} + \frac{\partial C}{\partial y} + \frac{\partial C}{\partial z} \right) \tag{2.5}$$

where is vapor flux, effective diffusivity, and concentration. D_{eff} is influenced by fabric porosity and tortuosity.

(ii) Capillary action drives liquid water within yarns and pores, described by the Lucas–Washburn equation:

$$L^2 = \frac{\gamma r \cos \theta}{2\mu} t \tag{2.6}$$

where is penetration length, surface tension, capillary radius, contact angle, and viscosity.

(iii) Moisture sorption–desorption occurs at the fiber level, governed by hygroscopic properties of cellulose or synthetic polymers. This process adds storage and release terms in the governing equations of coupled heat–mass transfer.

2.1.1.5 Dimensionless groups in simultaneous heat and mass transfer: Dimensionless groups or numbers are quantities without physical units that help to simplify complex physical phenomena by reducing variables. Several dimensionless numbers are often used in simultaneous hear and mass transport. These groups are essential for scaling lab drying studies to industrial machines like stenters and tunnel dryers.

(i) Biot number: The Biot number is the ratio of the internal resistance for conduction to the external resistance for convection of the body. It is given by:

$$Bi = \frac{hL_c}{k} \tag{2.7}$$

where L_c is the characteristic length.

(ii) Lewis number: It is defined as the ratio of thermal diffusivity to mass diffusivity and is given by:

$$Le = \frac{\propto}{D} \tag{2.8}$$

where $\propto$ is the thermal diffusivity and D is mass diffusivity.

(iii) Nusselt number: It is a dimensionless parameter used in convective heat transfer that represents the ratio of convective to conductive heat transfer.

$$Nu = \frac{hL}{k} \tag{2.9}$$

(iv) Sherwood number: It is a given by the ratio of convective mass transfer to mass transfer by molecular diffusion.

$$Sh = \frac{k_m L_c}{D} \tag{2.10}$$

where k_m is the convective mass transfer co-efficient.

2.1.2 Influence of Fabric Structure on Heat and Moisture Transport

The structural characteristics of fabrics, particularly porosity, pore size distribution, and permeability play a decisive role in governing the coupled heat and mass transfer processes occurring during textile drying, finishing, and heat-setting. A fabric can be modeled as a heterogeneous porous medium, where fibers form the solid matrix and air/moisture occupy the voids. The geometric configuration of this porous network determines the pathways for conduction, convection, and vapor diffusion, thereby influencing drying kinetics, energy efficiency, and final fabric quality.

2.1.2.1 Porosity (ε) is defined as the ratio of void volume to the total volume of the fabric.

$$\varepsilon = \frac{V_{void}}{V_{solid}} \tag{2.11}$$

Typical woven and knitted fabrics exhibit porosities in the range of 0.3–0.9, depending on yarn density, weave structure, and finishing treatments. Higher porosity generally enhances convective air penetration and moisture vapor diffusion but reduces solid-phase conduction, thereby lowering effective thermal conductivity.

Effective medium models are often used to relate porosity to thermal and mass transfer coefficients. For example, Maxwell–Eucken correlations provide an estimate of effective thermal conductivity [5, 32]:

$$\varepsilon = \frac{k_1 v_1 + k_2 v_2 \frac{3k_1}{2k_1 + k_2}}{v_1 + v_2 \frac{3k_1}{2k_1 + k_2}} \tag{2.12}$$

where k_1 and k_2 are the thermal conductivities of the fibers and air, respectively, and k_1 and k_2 are volume fractions of fibers and air respectively.

There are several other expressions available in literature which predicts effective thermal conductivity more accurately than Maxwell–Eucken correlations [9, 37].

2.1.2.2 Pore size distribution: The pore size distribution (PSD) characterizes the range and frequency of pore diameters within the fabric i.e. it represents the number of pores presents of different sizes. Textile materials often exhibit a bimodal PSD, with smaller pores within yarn bundles or intra-yarn pores and larger pores between yarns or inter-yarn pores. The PSD significantly affects capillary rise, drying rates, and vapor transport. The computational and experimental analysis has demonstrated that the heat transfer process is affected by the PSD of the porous textiles [38]. Experimentally, pore size distribution can be measured using mercury intrusion porosimetry, air permeability tests, and image-based analysis like SEM, FESEM etc. Advanced methods like X-ray microtomography provide 3D visualization of pore networks.

2.1.2.3 Permeability (B) quantifies the ease with which a fluid flows through the porous network of the fabric. Darcy's law is used to represent this phenomenon:

$$u = -\frac{B}{\mu} \Delta P \tag{2.13}$$

where u is the superficial velocity of air or liquid through the fabric, μ is the fluid viscosity, and ΔP is the pressure gradient across the fabric. Higher permeability enhances convective drying by allowing greater airflow through the fabric thickness, thus improving internal heat and mass transfer.

Kozeny–Carman equation, which was developed to provide a description of fluid flow in the packed bed can be used to calculate the permeability in porous media. One form of this equation which is commonly used for woven fabrics is given as [38]:

$$B = \frac{1}{K_0 S_0^2} \frac{\varepsilon^3}{(1 - \varepsilon)^2} \tag{2.14}$$

where K_0 is the Kozeny constant (typically ~ 5), and S_0 is the specific surface area of fibers. This shows that permeability increases strongly with porosity but decreases with higher surface area.

In convective dryers, air penetration depth is proportional to fabric permeability [11]. Low-permeability fabrics show surface drying and internal moisture retention, leading to uneven heat distribution. Conversely, highly permeable fabrics allow bulk convective exchange but may suffer from non-uniform shrinkage due to rapid surface evaporation.

These fundamental equations of mass (moisture) and heat transport in textiles (Eqs. 2.1–2.5) are highly complex to solve, particularly considering the structural properties of fibres. Thus their solutions are computationally demanding and solving them directly is often impractical. Therefore, researchers have developed a range of mathematical models empirical, semi-empirical, and lumped approaches [2, 8, 25, 26] to simplify analysis while retaining physical relevance. Studying these models in detail is essential to understand drying mechanism, optimize drying processes, minimize energy consumption [19], and advance sustainability by improving heat and mass transfer efficiency in textile drying.

2.1.3 Mechanism of Drying

Drying of textiles or other porous solids is mainly about removing water or solvent that is trapped inside the material. The process happens in two connected ways. First, the moisture inside the fibers or pores must move out to the surface. This movement depends on how easily water can travel through the material, which is controlled by the pore structure, fiber arrangement, and the strength of the bonds between water and the solid. In porous solids, liquid water can travel by four mechanisms [20]:

(i) Transport by capillary forces
(ii) Liquid diffusion
(iii) Pressure induced transport
(iv) Vapor diffusion.

Once the moisture reaches the surface, it is exposed to air or hot gases and evaporates into the surroundings.

This drying process is usually described in terms of two main stages after initial stage of preheating. At the beginning, there is a constant rate period where the surface of the material remains wet, and evaporation takes place steadily, much like water evaporating from an open surface. As drying continues, the surface becomes less wet and moisture must travel longer distances from inside the solid to the surface. This slows the process, giving the falling rate period, where the rate of drying decreases until the material reaches its equilibrium moisture content. Figure 2.1 shows different stages of drying.

Having outlined the fundamentals of heat and moisture transport in textiles, the next sections look at how these principles are applied in drying. Section 2.2 covers convection drying with heat recovery and control methods, Sect. 2.3 explains conduction drying with

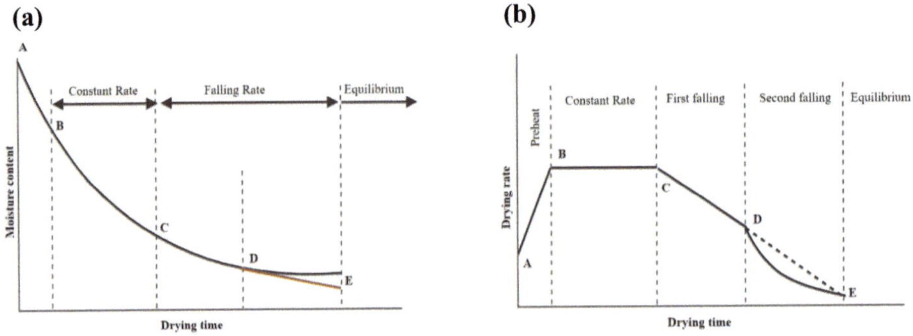

Fig. 2.1 Different stages of drying. **a** Moisture content and **b** drying rate

steam and renewable sources, and Sect. 2.4 discusses radiation-based methods such as infrared and microwave systems for better efficiency.

2.2 Convection Drying of Textiles

One of the most widespread industrial activities in textile finishing is drying, which most commonly employs convection [23]. It incorporates a heat exchange between the hot air (or any other gas medium) and the wet textile surface to allow water evaporation. Drying is accelerated by moving heated air around and through the textile, increasing heat and mass transfer rates. The basic working principle for this method is that the textile material is subjected to a hot air stream provided by blowers. Then it is further heated by gas, electricity, or steam. The moisture is moved from the fabric's interior into the surface and evaporates with the help of the thermal gradient. Variables like air flow rate, temperature, humidity, and other fabric characteristics like thickness and porosity influence the drying efficiency [14]. Convection drying is energy-intensive because it requires a lot of hot air, which needs to be heated and circulated. Nevertheless, a few eco-friendly alternatives can enhance its sustainability [6]. Figure 2.2 illustrates the classification of dryers based on operation.

2.2.1 Heat Recovery

The waste heat emitted from the exhaust air is recoverable and can be reutilized, significantly reducing the fuel utilization. Modern textile dryers widely make use of heat exchangers to perform this task [1]. It is seen that the use of heat recovery can result in reduced fuel consumption up to 28% in the restorative drying processes and 13.6% in

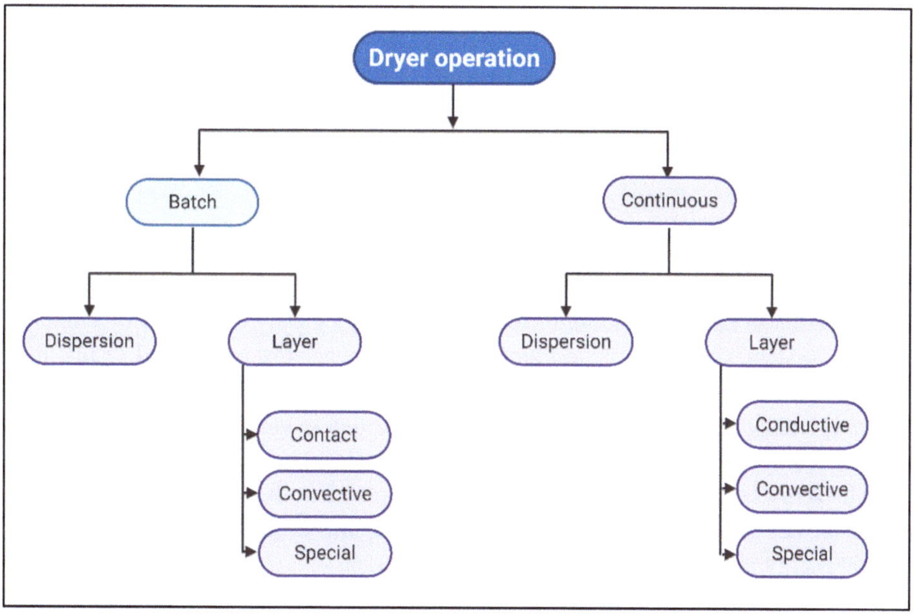

Fig. 2.2 Classification of dryers based on mode of operation

stenter machines [4, 6]. The heat recovery units can increase the energy efficiency by 59–87%, which depends on the type of material and the conditions during the operation [36]. According to the economic analysis of the heat recovery systems, these systems have low operating costs because they save fuel costs, which makes them a solution that the textile industry can use [16]. Also, as a result of heat recovery, there are reduced environmental effects, even when any type of fuel is chosen [36].

2.2.2 Moisture Sensors and Process Control

Installing state-of-the-art moisture sensors can ensure consistency in moisture on the coated surface to guarantee that the fabric is not over-dried. A combination of superior process control strategies, like the model-based control approaches with online moisture sensors, has indicated significant advantages in industrial uses. Such technologies give better control findings since they keep a careful eye and manage the moisture levels continuously, as seen in food processing plants [28].

With systems based on IoT, it is now possible to monitor and manage moisture in fabric materials in real time. They can continuously monitor moisture content during drying since these systems can be embedded in fabric. With the utilization of machine learning algorithms, such as K-Nearest Neighbors, it is possible to make predictive changes to drying conditions, to maximize energy consumption, and preserve the quality of the product [30]. The use of moisture sensors that have not been used earlier, like the ones using plasma polymerized films, is excellent in terms of resistance to water and chemicals, and the congruence of the given feeling concerning moisture. The sensors are vital in ensuring that the humidity in many industrial setups is well monitored [18].

2.2.3 Hybrid Systems

The reduction in the consumption of fossil fuels in drying can be achieved by employing a hybrid system, using convection alongside solar preheating or infrared technologies. These systems improve energy consumption efficiency and environmental sustainability by incorporating solar air collectors to warm the air before entering the dryer. Technologies such as hybrid drying systems have solar heat collectors, which preheat air, resulting in high energy savings as opposed to conventional practices. It is also seen that optimization of the design of solar air heaters and storage systems offers the least consumption of electric energy during different seasons [15].

Development of photovoltaic thermal (PV/T) systems uses the already heated air collected from a multi-direction solar collector, providing 35–52% efficiency, which reduces energy dependency on fossil fuels [21]. Intensive drying conditions go hand in hand with controlling the conditions of the drying, which can be maintained at an optimum drying temperature with hybrid innovative solar dryers being able to sharply cut drying time and energy consumption, with an energy-saving rate also reportedly between 25.54 and 77.1% in comparison with conventional processes [13]. By reducing fossil fuel dependency, these hybrid systems contribute to lower carbon emissions, promoting a more sustainable drying process.

2.3 Conduction Drying of Textiles

Another popular methods widely used in textile finishing is called conduction drying, or cylinder drying, which is direct heat transfer between hot surfaces and fabrics. The most common example is household iron press. This method is most useful for woven fabrics, as a smooth finish is ensured here through water evaporation by contact with the heated cylinder. The heat transfer and the mass transfer dynamics play a role in this process [17]. Conventional heat conduction drying systems have been highly energy-consuming and energy-dependent on fossil fuels. However, the latest technologies in terms of eco-friendly modalities have come up with sustainable measures, which go a long way in ensuring extensive energy efficiency. These systems are aimed at highly efficient steam control and waste heat utilization, which is essential to minimize environmental harm to the environment [4].

2.3.1 High-Efficiency Steam Management

Condensation recovery systems can be used to recycle condensed steam to be reused, instead of being wasted, resulting in greater efficiency. Additionally, appropriate insulations minimize heat loss during the transportation of steam, and hence, less energy is consumed when heating new environments. To reduce energy consumption, heat pumps can enable the recovery of the sensible and latent heat of exhaust gases, which could be used in drying. Unfortunately, not all systems are a reclaimed-heat-exhaust system because such a system can make significant savings of fuel up to 28% and can give efficiency increases of drying by 51% [4].

2.3.2 Use of Renewable Heating Sources

Implementing renewable heating through biomass boilers and solar thermal systems is another way to reduce the use of fossil fuels through drying in an industrial setting. In addition to energy efficiency, these technologies help ecological sustainability by reducing carbon emissions. Biomass boilers utilize organic materials, producing a renewable alternative to fossil fuels, accounting for 96% of renewable heat in the EU. Additionally, they can considerably decrease the emission of CO_2, making it more eco-friendly [33].

Combining biomass with solar thermal energy can increase efficiency, save a lot of biomass fuel, and promote efficient boilers [27]. The biomass power generator hybrid solar thermal power system is used to enhance the temperature of feedwater and primary air temperature using solar energy, which will make the boiler efficient and decrease the dependence on fossil fuel. This integration will suitably enable drying processes whilst increasing the utilization of renewable energy in thermal systems.

Solar thermal technology can be applied successfully in industrial processes and replace fossil fuels. For example, a linear Fresnel system proved to be less costly and has a smaller carbon footprint than conventional approaches [34]. Solar thermal systems combined with biomass have the potential to maximize energy buys so that heating facilities maintain the same energy production without harming the environment [24].

2.4 Radiation Drying of Textiles

Textile radiation drying employs electromagnetic waves, such as infrared, microwave, and radio-frequency, to impart heat more effectively than conventional means. Infrared radiation penetrates the surface and heats it for direct heating, whereas microwaves give volumetric heating through the agitation of water molecules in the material. This unique concept increases the drying capacity and quality of products. This drying technology is significantly eco-friendlier compared to other methods. Radiation drying can quickly and uniformly dry thick or high-moisture fabric because it can be done in more space- and energy-efficient systems. These are very small in size, facilitating fitting into continuous lines [12].

2.4.1 Infrared with Renewable Power

Infrared drying (IRD), especially combined with alternative, renewable energy sources, is one solution that can potentially increase the energy efficiency and sustainability of drying. This way, not only is drying much faster, but energy usage and environmental burden are also much smaller. Infrared radiation heats the surface of the textile directly, causing a quick evaporation of the moisture. This technique is also found to reduce drying periods greatly [12].

IRD technologies, like Catalytic IR, have far better drying power than older systems and lead to reduced energy consumption and avoidance of waste. Experimental works indicate that incorporating photovoltaic-thermal pairing can increase energy savings, making the process more sustainable. Exploiting the sources of energy that can be replenished, like solar power, in infrared drying systems helps dent the amount of carbon emissions and decreases dependence on fossil fuels. The incorporation of infrared technology with solar power has not only saved energy but also ensured exemplary quality drying achievements, which is essential in the case of agricultural products [10].

Although infrared drying using renewable energy resources is associated with many benefits, challenges such as the high cost of addressing initial expenses and the requirement to provide appropriate processing conditions have to be resolved to ensure the full potential of the current technology in different industries.

2.4.2 Microwave-Assisted Drying

Microwave heating applies volumetric heating and interacts with water molecules to heat the material uniformly to remove moisture. This has been proven to boost the performance of textile processing [10]. Microwave-assisted drying (MAD) of fabrics offers a potential substitute method for conventional drying procedures, focusing on energy conservation and minimising environmental hazards. The advantage of using this technology is that it dries fast while using minimum energy and pollutant emissions. MAD dries the textiles much quicker than conventional techniques, and this significantly decreases energy consumption [31].

Combining the MAD with systems like fluidised bed dryers with multi-stage microwave heating can recycle heat and minimise energy wastage. MAD cuts the emission of volatile organic compounds by utilizing condensing exhaust that collects the materials to be treated, hence addressing concerns about pollution. The application of reduced pressure in the drying operations results in a lower boiling point of water, which also boosts energy efficiency and has fewer adverse effects on the environment.

The constant research aims to enhance the stability and safety of the MAD systems to enable their full deployment in the textile manufacturing sector. Although MAD has many benefits, technological barriers and the requirement of acceptance on an industry-wide basis are issues of concern. To remedy these concerns, they are of vital importance in the deployment of sustainable drying technology in the textile process [31].

2.4.3 Hybrid Radiation-Convection Systems

The optimization of hybrid systems and radiation-convection systems in drying has highlighted the promotion of efficiency and the minimization of energy consumed. Such systems make it possible to combine surface radiation heat with deep moisture drainage convection to control drying dynamics and energy consumption. Hybrid systems have proved to reduce the drying time. In a study, drying time was reduced by 40% compared to the conventional means, and a moisture reduction rate of 0.8% per minute showed the effectiveness of hybrid systems [7]. Increasing the level of radiation and maintaining an appropriate amount of airflow concurrently can optimize moisture removal; hence, a faster rate of drying was observed when the radiation level was raised.

The effect of convection combined with radiation has been associated with high savings in energy. Optimizing radiation intensity and airflow led to a reduction of energy consumption in the process of garlic slice drying by 36% [29]. The hybrid design also reduces the generation of greenhouse gases, where solutions such as the catalytic heater result in highly efficient combustion systems that have low emission levels [7]. Experimental data suggest an enhanced mass transfer rate by the hybrid-drying method, which results in uniform temperatures throughout the material and effective moisture removal across several materials, including granular materials and bee pollen. Even though hybrid systems offer many advantages, they pose a challenge in developing ideal operation parameters to eliminate the possibility of heightened energy losses that occur due to transient effects in moisture ratios and drying rates. The systems require further research to make them relevant to wider applications [3]. A comparative summary of the energy-saving potential and efficiency improvements reported across various eco-friendly drying techniques is presented in Table 2.1.

Table 2.1 Energy savings and efficiency improvements in eco-friendly textile drying

Drying method	Eco-friendly approach	Efficiency and savings
Convection drying	Heat recovery (exhaust reuse in stenter/restorative drying)	Fuel savings up to 28%; 13.6% in stenter machines; energy efficiency increase 59–87%
	Moisture sensors and process control (IoT, ML-based)	Improved quality; reduced over-drying; optimized energy use (qualitative)
	Hybrid solar/IR-assisted drying	Energy savings 25.5–77.1% compared to conventional methods; PV/T efficiency 35–52%
Conduction drying	Steam condensate recovery and insulation	Fuel savings up to 28%; drying efficiency increases by 51%
	Biomass boilers (EU)	Accounts for 96% of renewable heat; significant CO_2 emission reduction
	Solar thermal + biomass hybrid	Higher efficiency; lower biomass use; reduced fossil fuel dependence
Radiation drying	Infrared drying (with renewables)	Reduced drying time; significant energy savings (qualitative); higher efficiency with PV/T integration
	Microwave-assisted drying (MAD)	Faster drying; significant energy savings; VOC emission reduction; energy use cut substantially
	Hybrid radiation–convection	Drying time reduced by 40%; energy consumption cut by 36%

2.5 Concluding Remarks

The textile sector has become one of the most resource-consuming industries in the world, which leads to the need to redefine the balance between industrial efficiency and environmental sustainability, which is currently at a critical stage. Drying is one of the most energy-intensive procedures of any textile manufacturing company due to the energy expenditure, which amounts to almost a quarter of the aggregate energy spent by the company. Historically, textile drying has been based on convection, conduction, and radiation, which are effective in moisture elimination, but not sustainable as they require fossil fuels to generate energy which results in greenhouse gases. The analysis of drying indicates a possible change with the engagement of technological advancement, the use of renewable energy, and even during the optimization of this process. One of the most popular drying methods, convection drying, is very energy-demanding in terms of heated air circulation. Waste heat recovery, improved moisture detection, and hybrid convection systems made of solar and photovoltaic thermal (PV/T) show promising potential to curtail operational expenditure and carbon emissions. The innovations save on energy consumption and take textiles to a higher quality by providing uninterrupted drying and avoiding over-processing. The implementation of IoT-based monitoring and predictive analytics is also a future evidence of the transition to the development of so-called innovative drying systems that can constantly adjust to ever-changing conditions of the fabrics in use, allowing drying systems to become more efficient and more sustainable.

Conduction drying, which is widely used through cylinder dryers, also experiences the issue of wastage of energy. Nevertheless, with highly efficient steam management systems (such as heat pumps and condensation recovery systems), efficiency has been shown to reduce wastes considerably. The shift from steam-based drying using fossil fuels to renewable resources like biomass and solar thermal heat improves the green perspective of conduction drying. Relevantly, hybrid biomass-solar systems not only lead to a reduction of carbon footprint, but they also increase energy supply reliability to textile industries located in various geographic regions and applicable climatic conditions. The combination of renewable sources makes conduction drying sustainable and aligned with the world's sustainability agenda.

Radiation drying, in its turn, is the most advanced area of sustainable textile finishing. Drying fueled by renewable energy, microwave-assisted drying, and combined radiation convective drying all reported excellent improvements in drying time, energy use, and air pollution, respectively. Microwave and radio frequency heating is volumetric, providing consistent drying even for thick and 3D fabrics with high-moisture content. The hybrid technology design better utilizes the advantages of more than one drying unit to enhance efficiency. With up to 77% energy savings, these technologies offer a scalable and future-ready alternative to current methodologies. Nonetheless, the high cost of initial investments and the technological barrier to adaptation may also be obstacles that the industry can only handle with the help of additional research, policies, and joint innovation. The combination of all these developments indicates a shift in paradigm in the field of textile processing, where sustainability is no longer a secondary but a primary design factor. Eco-friendly drying technologies are part of the broader initiative to reduce carbon contribution, save natural resources, and enhance the quality of the products in terms of low prices. In the case of the textile industry, such green practices are part of a long-term sustainable strategy to withstand competition in a global market. In the future, renewable energy integration, digital process control, and hybrid technological models will establish the next generation of textile drying systems, thus making sure that this vital part of textile production becomes environmentally friendly, financially viable, and technologically advanced.

References

1. Acar, C., Dincer, I., & Mujumdar, A. (2022). A comprehensive review of recent advances in renewable-based drying technologies for a sustainable future. *Drying Technology*, *40*(6), 1029–1050. https://doi.org/10.1080/07373937.2020.1848858

2. Akyol, U., Erhan Akan, A., & Durak, A. (2015). Simulation and thermodynamic analysis of a hot-air textile drying process. *The Journal of The Textile Institute, 106*(3), 260–274. https://doi.org/10.1080/00405000.2014.916062
3. Brandão, R. J., Santos, L. D. M., Marques, L. G., & Prado, M. M. (2015). Heat and Mass Transfer and Energy Aspects in Combined Infrared-Convective Drying of Bee-Pollen. *Defect and Diffusion Forum, 364*, 9–17. https://doi.org/10.4028/www.scientific.net/DDF.364.9
4. Budin, Sc. R., & Mihelić-Bogdanić, Sc. A. (2011). Reducing Energy Supply In The Drying Process. *Energy Engineering, 108*(2), 6–16. https://doi.org/10.1080/01998595.2011.10389015
5. Carotenuto, G. B. A., (2000). The effective thermal conductivity of packed beds of spheres for a finite contact area. *Numerical Heat Transfer: Part A: Applications, 37*(4), 343–357.
6. Cinar, G., & Ceylan, H. (2021). The effect of the heat recovery on fuel consumption in the stenter machine. *Thermal Science, 25*(2 Part A), 1047–1055. https://doi.org/10.2298/TSCI190510336C
7. Ebrahimpour, V., & Esmailpour, K. (2024). Simultaneous use of convection and radiation heat transfer for Iranian pistachio drying using catalytic heater—Numerical and experimental study. *Frontiers in Mechanical Engineering, 10*, 1459806. https://doi.org/10.3389/fmech.2024.1459806
8. Fontana, É., Donca, R., Mancusi, E., Ulson De Souza, A. A., & Guelli Ulson De Souza S. M. A., (2016). Mathematical modeling and numerical simulation of heat and moisture transfer in a porous textile medium. *The Journal of The Textile Institute, 107*(5), 672–682. https://doi.org/10.1080/00405000.2015.1061324
9. Gonzo, E. E. (2002). Estimating correlations for the effective thermal conductivity of granular materials. *Chemical Engineering Journal, 90*(3), 299–302. https://doi.org/10.1016/S1385-8947(02)00121-3.
10. Griffin, W. L., & Hendrix, W. A. (1986). Microwave Heating and Drying in Textile Processing-Present and Future. *IEEE Transactions on Industry Applications, IA-22*(1), 115–120. https://doi.org/10.1109/TIA.1986.4504691
11. Haghi, A. K. (2011). *Heat and mass transfer in textiles* (2nd ed). WSEAS Press.
12. Hay, N., Quang Huy, L., & Van Kien, P. (2023). A Review of Drying Methods Assisted by Infrared Radiation, Microwave and Radio Frequency. In J. D. Jovanovic (Ed.), *A Comprehensive Review of the Versatile Dehydration Processes*. IntechOpen. https://doi.org/10.5772/intechopen.108650
13. Ibrahim, A., Elsebaee, I., Amer, A., Aboelasaad, G., El-Bediwy, A., & El-Kholy, M. (2023). Development and evaluation of a hybrid smart solar dryer. *Journal of Food Science, 88*(9), 3859–3878. https://doi.org/10.1111/1750-3841.16713
14. Johann, G., Silva, E. A., Lima, O. C. M., & Pereira, N. C. (2014). Mathematical modeling of a convective textile drying process. *Brazilian Journal of Chemical Engineering, 31*(4), 959–965. https://doi.org/10.1590/0104-6632.20140314s00002685
15. Khattab, N. M. (1996). Optimization of Hybrid Solar Dryer. *Energy Sources, 18*(7), 781–790. https://doi.org/10.1080/00908319608908810
16. Kim, J., Mun, H., Shim, J. Y., Lee, I., & Cho, H. (2024). Advanced energy recovery systems design of stenter processes: Energy, exergy and Techno-economic analyses. *Energy, 289*, 130028. https://doi.org/10.1016/j.energy.2023.130028
17. Krasnikov, V. V. (1970). Contact heat transfer and external mass transfer in contact drying. *Journal of Engineering Physics, 19*(3), 1136–1142. https://doi.org/10.1007/BF00826238
18. Kraus, F., Cruz, S., & Müller, J. (2003). Plasmapolymerized silicon organic thin films from HMDSN for capacitive humidity sensors. *Sensors and Actuators B: Chemical, 88*(3), 300–311. https://doi.org/10.1016/S0925-4005(02)00373-8.

19. Lambert, A. J. D., Spruit, F. P. M., & Claus, J. (1991). Modelling as a tool for evaluating the effects of energy-saving measures. Case study: A tumbler drier. *Applied Energy, 38*(1), 33–47. https://doi.org/10.1016/0306-2619(91)90040-5.
20. McCabe, W. L., Smith, J. C., & Harriott, P. (1993). *Unit operations of chemical engineering*. McGraw-Hill.
21. Mohd Azmi, M. S., Abdul Majid, Z. A., & Ruslan, M. H. (2022). DEVELOPMENT AND PERFORMANCE OF A SOLAR DRYER HYBRID PHOTOVOLTAIC THERMAL SYSTEM: PEMBANGUNAN DAN PRESTASI SISTEM PENGERING SURIA HIBRID FOTOVOLTAN TERMA. *Jurnal Teknologi, 85*(1), 53–61. https://doi.org/10.11113/jurnalteknologi.v85.14845
22. Mujumdar, A. S. (2001). Industrial Drying Technologies: Current Status and Future Trends. In Y. H. Mori & K. Ohnishi (Eds.), *Energy and Environment* (pp. 112–125). Springer Japan. https://doi.org/10.1007/978-4-431-68325-4_5
23. Mujumdar, A. S. (Ed.). (2014). *Handbook of Industrial Drying* (0 ed.). CRC Press. https://doi.org/10.1201/b17208
24. Mutlu, Ö. Ç., Büchner, D., Theurich, S., & Zeng, T. (2021). Combined Use of Solar and Biomass Energy for Sustainable and Cost-Effective Low-Temperature Drying of Food Processing Residues on Industrial-Scale. *Energies, 14*(3), 561. https://doi.org/10.3390/en14030561
25. Ng, A. B., & Deng, S. (2008). A new termination control method for a clothes drying process in a clothes dryer. *Applied Energy, 85*(9), 818–829. https://doi.org/10.1016/j.apenergy.2007.10.016
26. Nordon, P., & David, H. G. (1967). Coupled diffusion of moisture and heat in hygroscopic textile materials. *International Journal of Heat and Mass Transfer, 10*(7), 853–866. https://doi.org/10.1016/0017-9310(67)90065-8.
27. Okoroigwe, E. C., Eke, M. N., & Ugwu, H. U. (2013). Design and evaluation of combined solar and biomass dryer for small and medium enterprises for developing countries. *International Journal of Physical Sciences, 8*(25), 1341–1349. https://doi.org/10.5897/IJPS2013.3937
28. Porciello, G. P., & Doerr, D. (1999). Advanced process control for moisture monitoring and control applications. *IEEE Industry Applications Society Advanced Process Control Applications for Industry Workshop*, 58–64. https://doi.org/10.1109/APCA.1999.805028
29. Qenawy, M., Ali, M., El-Mesery, H. S., & Hu, Z. (2024). Analysis and control of hybrid convection-radiation drying systems toward energy saving strategy. *Journal of Food Science, 89*(12), 9559–9576. https://doi.org/10.1111/1750-3841.17522
30. Ranganathan, C. S., Riazulhameed, A. A. M. A., Saravanan, S., Mohankumar, N., Rajmohan, M., & Murugan, S. (2024). IoT Based Real-Time Moisture Management in Fabric Drying Processes. *2024 First International Conference on Innovations in Communications, Electrical and Computer Engineering (ICICEC)*, 1–6. https://doi.org/10.1109/ICICEC62498.2024.10808630
31. Rattanadecho, P., & Makul, N. (2016). Microwave-Assisted Drying: A Review of the State-of-the-Art. *Drying Technology, 34*(1), 1–38. https://doi.org/10.1080/07373937.2014.957764
32. Rocha, R. P., & Cruz, M. A. E. (2001). Computation of the effective conductivity of unidirectional fibrous composites with an interfacial thermal resistance. *Numerical Heat Transfer: Part A: Applications, 39*(2), 179–203.
33. Saidur, R., Abdelaziz, E. A., Demirbas, A., Hossain, M. S., & Mekhilef, S. (2011). A review on biomass as a fuel for boilers. *Renewable and Sustainable Energy Reviews, 15*(5), 2262–2289. https://doi.org/10.1016/j.rser.2011.02.015
34. Seyitini, L., & Enweremadu, C. C. (2024). Assessing the Applicability of Solar Thermal Technologies for Industrial Tea Drying. *CONECT. International Scientific Conference of Environmental and Climate Technologies*, 82. https://doi.org/10.7250/CONECT.2024.058

35. Syeda Mehdia Naqvi, Arfan Bakhtiar, & Manik Mahachandra. (2024). Energy audit of a textile company: A case study of finishing textile. *World Journal of Advanced Research and Reviews*, *22*(3), 1040–1047. https://doi.org/10.30574/wjarr.2024.22.3.1538
36. Urbano, D. G., Aquino, A., & Scrucca, F. (2023). Energy Performance, Environmental Impacts and Costs of a Drying System: Life Cycle Analysis of Conventional and Heat Recovery Scenarios. *Energies*, *16*(3), 1523. https://doi.org/10.3390/en16031523
37. Wang, J., Carson, J. K., North, M. F., & Cleland, D. J. (2006). A new approach to modelling the effective thermal conductivity of heterogeneous materials. *International Journal of Heat and Mass Transfer*, *49*(17–18), 3075–3083. https://doi.org/10.1016/j.ijheatmasstransfer.2006.02.007
38. Zhu, Q., & Li, Y. (2003). Effects of pore size distribution and fiber diameter on the coupled heat and liquid moisture transfer in porous textiles. *International Journal of Heat and Mass Transfer*, *46*(26), 5099–5111. https://doi.org/10.1016/S0017-9310(03)00370-3.

Eco-Friendly Chemical Treatment Methods 3

Contents

Abstract

The textile industry is very resource and chemical-demanding; wet processes, including pretreatment, dyeing, printing, and finishing, are significant contributors to energy consumption, chemical waste, and water contamination. The conventional techniques involve permanent substances, high liquor ratios, and toxic auxiliaries, all of which have negative effects on the environment and human health. Green chemistry uses alternatives that are driven by degradability, prevention, and green solvents to replace solvents. Enzyme-assisted pretreatments, salt-free or supercritical CO_2 dyeing, plasma

A. H. Bari et al., *Eco-Friendly Textile Processes*, Synthesis Lectures on Chemical Engineering and Biochemical Engineering, https://doi.org/10.1007/978-3-032-08916-8_3

treatment, ozone/UV oxidation, the use of natural dyes, and bio-mordants are examples of ecological substitutes for the earlier noted methods. Modern methods, including sol-gel coatings, layer-by-layer assembly, and nanoparticle finishing, reduce effluents while adding antibacterial, UV protective, water-repellent, and self-cleaning properties. Without sacrificing comfort, nanotechnology and plasma will improve multifunctionality, durability, and wash-fastness. One of the traditional uses of circular frameworks is dabu printing. Combining biotechnology, nanotechnology, and renewable chemistries enables the production of high-performance, environmentally friendly textiles that are safe and promote the objectives of the circular economy.

Keywords

Sustainability • Green chemistry • Enzyme-based finishing • Plasma treatments • Nanotechnology

3.1 Introduction

The modern fabric industry uses a lot of wet-processing, which is very chemically and water-intensive and includes pretreatment, dyeing/printing, and finishing. Conventional techniques generate effluents with complicated chemical characteristics and high color, use a lot of thermal and electrical energy, and consume considerable amounts of water and auxiliaries (salts, alkalis, surfactants, and dispersants). These effluents frequently contain the release of bioactive, persistent, and even poisonous compounds that can endanger aquatic ecosystems and human health while also placing strain on wastewater treatment facilities. Wet-processing is frequently mentioned in most recent assessments as one of the major environmental hotspots in textile supply chains, particularly in cotton and polyester fabric. This has prompted a shift to so-called green or eco-friendly treatment methods that use less energy, water, and dangerous chemicals while maintaining or even improving product performance [7, 66].

Traditional dyeing and printing methods typically involve high liquor ratios, numerous rinses, and heat treatments. As a result, it significantly affects the amount of freshwater that is drawn down and the amount of heat energy needed for each garment. The end effluents are difficult to treat and degrade receiving waters because of color, chemical oxygen demand, salts, and residual auxiliaries. Many synthetic dyes are also resistant to biological degradation. Critical evaluations of textile effluents demonstrate color persistence, the phenomenon of certain azo dyes forming aromatic amines, and the incapacity to eliminate complexed metals and finishing chemicals [46]. Green chemistry provides a conceptual underlying for re-engineering textile processing. Textile wet-processing pain points are directly addressed by the twelve principles, which translate to prevention, atom economy, safer solvents and auxiliaries, energy efficiency, catalysis, design for degradation, and inherently safer chemistry (e.g., solvent/water use, auxiliary load, high-temperature steps,

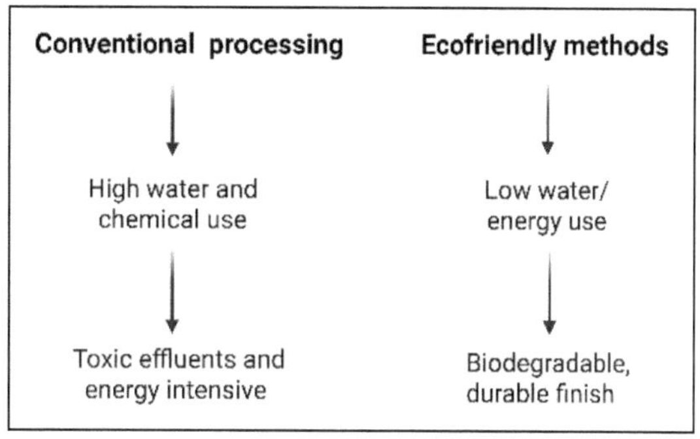

Fig. 3.1 Comparison of conventional and eco-friendly methods

and the persistence of residues). Applying these concepts can result in the substitution of media (foam/inkjet), the reduction of dyeing temperatures through catalysis or plasma activation, the replacement of harsh alkalinity bleaching by enzymes, and the preference for degradable dyes/auxiliaries made from sustainable feedstocks [8, 26]. Figure 3.1 compares the conventional and eco-friendly methods.

The successful application of green chemistry principles is growing, with studies concentrating on tangible instances in the textile industry. Enzyme-based pretreatments offer a softer environment and less fiber destruction and replacing strong alkaline scouring and oxidative bleaching. The low-additive dyeing techniques lower the auxiliary charges and effluent toxicity without changing the shade level or depth. Solvent-free surface activation techniques, such as those used in plasma or ozone processing, use less water and provide functional finishes with few residues. Similarly, non-hazardous sources are non-toxic or renewable resins that are substitutes for formaldehyde-based resins [42]. In order to achieve high dye uptake, exceptional leveling, and the avoidance of dispersants and washing, supercritical CO_2 dyeing uses CO_2 as a transport medium instead of water. This results in a large decrease in water use and effluent production and often in total energy per kg dyed (no post-drying). Advances in polyester to other fibers/blends and the integration of functionalization processes are documented in recent reviews and studies. The environmental benefits are apparent where product mix application is acceptable, despite the fact that capital intensity and dye compatibility continue to be the obstacles [11].

Low-temperature plasma often replaces or increases the effectiveness of wet pick-ups by changing only the outermost layers of fibers, either by improving wettability/adhesion or by depositing incredibly thin functional films. Because plasma is a dry process, it uses less water and auxiliary water, and it can shorten the time or temperature needed for curing. Recent reviews and case studies demonstrate plasma-based improvements in printability and image quality, antibacterial qualities, and the adsorption of cotton, polyester, and silk dyes, as well as quantifiable savings in processing resources [64].

3.2 Chemical Treatment Principles: Functionalization and Finishing

Functionalization, or adding specific, valuable features to fibers or fabrics, and finishing, which improves mechanical, aesthetic, or comfort-based performance, are the two main areas into which textile chemical treatment is typically separated. Functionalization can add performance qualities like flame resistance, UV block, water repellency, antimicrobial activity; finishing can add qualities like handling, softness, wrinkle-free, or stable dimensions. These treatments also make the textile easier to work with and extend its service life. The most prevalent solutions are based on formaldehyde-based cross-linkers, halogenated compounds, or synthetic resins, which produce long-lasting pollutants and toxic effluent. The idea behind eco-friendly functionalization is that fibers and finishes must interact strongly (via plasma-activated covalent bonds, ionic bonds, or hydrogen bonds) to reduce the need for repeated finishes and, consequently, the chemical footprint. Layer-by-layer assembly, cyclodextrin-based host carry to, and enzymatic grafting studies are other viable, sustainable source functionalization techniques [17, 20].

3.2.1 Functionalization—Building New Performance into Fibers/Fabrics

In textiles, functionalization is commonly defined as adding new functions to the bulk or surface of the fiber without significantly altering its handling or aesthetic qualities. It is widely used to refer to the provision of new, advanced, long-lasting, and sustainable performance. To offer antibacterial and antiviral activity, biocidal chemistries such as quaternary ammonium, metal/metal-oxide nanoparticles (Ag, Cu, ZnO), or bio-active compounds can be grafted or deposited; plasma and sol-gel deposition allow for improved anchoring. Silicone, hydrocarbon, and hybrid sol-gel coatings are used in fluorine-free water and oil-repellent finishes. Low-surface energy treatments and nanofiber surfaces that provide breathable protection are used to enhance hierarchical roughness [20]. Functionalization refers to the deliberate, regulated addition of new qualities to the fabric surface and frequently the near-surface bulk without a noticeable loss of hand or drape. Common techniques include layer-by-layer for producing adjustable ultrathin deposits, sol-gel networks as long-term durable skins, functionalization by plasma treatment to provide reactive groups and improve adhesion, and covalent grafting of polymers. Together, they provide a powerful fixation of the active ingredients and are resistant to wear and washing [62]. In order to make textiles a competitive alternative in high-tech and sustainable markets, functionalization techniques are used to produce durability (wash-fastness), safety (eco-friendly chemistry), and multifunctionality [40].

Antimicrobial and antiviral textiles have gained popularity, particularly in the medical and protective fields, since they assist hygienic security. In order to destroy microbes, they are typically introduced by encapsulating silver, copper, or zinc oxide nanoparticles, which release biocidal ions or produce reactive oxygen species. Quaternary ammonium salt-based ones provide killing when they come into contact with a surface, but photocatalytic surfaces like TiO_2 self-sterilize when exposed to light. These functionalizations are essential for medical textiles, personal protective equipment, and intelligent healthcare materials because they not only improve the resistance of fabric to germs and viruses, but they also extend the length of protection and allow for greater fabric reuse [55]. According to recent developments, there are two methods for using nanoparticles to make textiles hydrophobic and UV-protected. Because zinc oxide (ZnO) and titanium dioxide (TiO_2) nanoparticles strongly absorb and scatter harmful UV rays, they are effective UV blockers that improve wearer safety and fabric durability. Together, carbon allotropes offer multifunctional performance and photostability. Since hydrophobic coatings that are fluorine-free demonstrate long-term maintained water repellency without being environmentally persistent, they will be employed to lessen the sustainability element. Performance and safety are balanced by this sustainable approach, which can also produce breathable, moisture-resistant textiles with longer useful lives [53].

Smart textiles are rapidly incorporating emerging nanomaterials such as MXenes, graphene, and carbon nanotubes. These materials give fabrics that can detect physiological signals, warm the surrounding region, and generate electrical energy when movements are performed or ambient power is captured very good electrical conductivity, stretchability, and wearability. Such multipurpose fabrics' performance is particularly beneficial in wearable electronics and healthcare monitoring, where success depends critically on their ability to be both lightweight and comfortable [9]. Recent developments in the field of textile functionalization have placed a greater focus on environmental concerns, with the goal of reducing the negative environmental effects of goods while enhancing their functionality. Since enzyme-assisted pretreatments replace harsh chemicals and produce less wastewater, they are more energy-efficient and selective surface modification techniques. An alternative to heat methods, plasma curing is a low-energy, waterless method that offers a more robust attachment of functional coatings without consuming excessive amounts of resources. Furthermore, because the developed PFAS-free water repellents eliminate harmful, persistent, and ubiquitous fluorochemicals while retaining their water-repellent properties, they meet today's sustainability requirements [45] (Table 3.1).

Plasma and Surface Activation

The processes of surface activation and plasma activation include the usage of non-equilibrium plasma, which interacts with the textile fiber surfaces and forms various energy species. Reactive functional groups (hydroxyl and carboxyl) are added to the fiber surface by a mechanism that significantly increases surface energy and guarantees good wettability, dyeability, and adherence of the applied coating or finishing. Plasma treatment is a clean and efficient method of functionalizing textiles since it only modifies the

Table 3.1 Functionalization techniques and their applications

Technique	Materials used	Functional properties	Application area
Plasma activation	O_2, air plasma	Improved wettability, dyeability	Cotton, polyester fabrics
Sol–gel coatings	Hybrid organic–inorganic sols	Antimicrobial, UV-blocking, flame-retardant	Technical textiles
Layer-by-layer assembly	Polyelectrolytes, nanoparticles	Self-cleaning, flame retardant, conductive	Smart/protective textiles
Nanoparticle deposition	ZnO, TiO_2, Ag NPs	Antimicrobial, UV protection	Healthcare, sportswear
Enzymatic grafting	Laccase, peroxidase	Antimicrobial, eco-friendly finishes	Cellulosic fabrics

surface, preventing bulk damage to the material and ineffective base material penetration [48].

Sol–Gel Coatings

Hybrid organic-inorganic sol-gel networks are becoming more and more popular in textile finishing because they offer the chemical and thermal stability of inorganic oxides along with the flexibility of organic polymers. During use or instrument cleaning, they are held in place by a stable fiber surface network that can stabilize hydrophobic chemicals, flame retardants, and nanoparticles. In the long run, this longevity improves the functionality of functional coatings. Additionally, by allowing for multifunctionality, the sol-gel process makes it possible to incorporate flame-resistant, UV-protective, antimicrobial, water-repellent, or combinations of these properties into a single textile without compromising its handle, appearance, or breathability. This makes the textile incredibly versatile in its advanced capabilities [62].

Layer-by-Layer Assembly

Layer-by-layer assembly is a general surface modification technique that creates highly ordered multilayers by alternating the deposition of oppositely charged polyelectrolytes, nanoparticles, or nanomaterials. The method is highly appropriate for advanced textile coating since it offers nanometer-scale control over layer thickness, composition, and functionality. Textiles can be made to be antimicrobial, self-cleaning, flame-retardant, or electrically conductive by varying the number of layers and the kind of material used. In order to preserve the inherent softness and flexibility of textiles and provide a multi-functional performance that meets the demands of intelligent and protective textiles, the treatment is water-based, extremely delicate, and appropriate for a variety of fiber types [31, 41].

3.2.2 Finishing—Tuning Comfort, Handle, and Serviceability

In the textile industry, finishing refers to the step of applying a product that includes additional treatments after the fabric has been formed and dyed to improve comfort, handling, durability, and serviceability. On the other hand, finishing is mostly used to enhance the usefulness and beauty of textiles without adding any new qualities. Common finishing processes that provide a pleasant combination of performance include chemical finishing (softeners, resins, cross-linkers), mechanical finishing, and environmentally friendly enzyme treatment [37]. Chemical softeners, including silicones, fatty acid derivatives, and newly developed bio-based softeners, are frequently used to enhance the softness, drape, and flexibility of textiles without compromising their durability. The individual who receives these treatments feels more at ease because their hands feel smooth. Enzymatic bio-polishing is a useful addition to this process since it efficiently eliminates surface uncertainty and projecting fibers, giving the finish a glossier, cleaner appearance with less pilling. Since they don't seem as appealing and sustainable, these sustainable techniques collectively encourage more useful and appealing forms [29].

Dimensional stability and rip resistance are two finishing requirements in textile manufacturing that help maintain shape and beauty after washing and wearing. Durable-press and anti-crease properties were formerly given by formaldehyde-based resins, but safer substitutes are being investigated because of the health and environmental consequences. Recent developments have employed citric acid and polycarboxylic acids as crosslinkers, which are safe for the environment and readily bind to cellulose via an ester bond. These treatments improve crease recovery, reduce shrinkage, reduce mechanical strength, and offer long-lasting remedies [43]. High-performance textiles need to be serviceable and have adequate moisture management. Hydrophilic finishes are crucial for supporting high-performance sportswear during activity because they significantly improve sweat absorption and rapid evaporation rates. Additionally, because dirt and stains are not embedded in the fibers, soil-release and stain-resistant treatments enhance cleaning capabilities. These finishes prolong the functional life of clothing, reduce cleaning, and preserve its attractiveness. All of them work together to produce practical, sustainable textiles [18, 67].

Green chemistry concepts are becoming more significant as eco-friendly advancements in textile finishing take place. Bio-based crosslinkers or enzyme-based treatments are replacing plasma procedures or fluorochemical-free softeners as environmentally acceptable alternatives to traditional techniques that employ formaldehyde-based resins and fluorochemical-based crosslinks. These solutions reduce the use of harmful chemicals, have less of an impact on the environment, and satisfy consumer needs for safer and more ecologically friendly fabrics. Using such eco-friendly technology not only ensures improved performance but also increases the long-term sustainability of the textile industry [45] (Table 3.2).

Table 3.2 Sustainable finishing treatments

Technique	Eco-friendly approach	Properties added	Environmental benefit
Enzyme-based finishing	Cellulase, catalase, laccase	Bio-polishing, antibacterial	Low energy, reduced COD/BOD
Plasma finishing	Low-temperature/ atmospheric plasma	Hydrophobicity, self-cleaning	Waterless, chemical-free
UV/ozone finishing	Radical-based oxidation	Antimicrobial, bleaching	Zero effluent, low energy.
Nanotechnology finishing	ZnO, TiO_2, chitosan NPs	Antibacterial, UV-blocking	Renewable, multifunctional
Bio-based softeners	Chitosan, fatty acid derivatives	Soft hand feel, drape	Non-toxic, biodegradable

3.3 Eco-Friendly Dyeing and Printing Techniques

Over the past few years, a rising number of textile companies have implemented eco-friendly textile printing and dying methods in an effort to reduce the amount of water and chemicals needed, as well as to increase color fixation and reduce energy usage. One noteworthy technique is salt-free reactive dyeing; for example, cotton textiles were effectively altered using graft polymerization with chitosan and renewable cross-linkers such dimethyl itaconate without losing color fastness or depth of color. Additionally, salt-free dyeing for reactive and direct dyes would be possible using biopolymer mordants like chitosan, keratin, and others. These mordants would also offer a suitable degree of fixing, wash fastness, and even an antibacterial impact [36]. Another alternative process, solvent-aided salt-free dyeing, does away with water as a solvent, substituting it instead with eco-friendly liquid solvents such as ethanol and isopropanol, achieving much greater color strength exhaustion, and fixation than in aqueous dyeing processes, again with superior fastness levels as well [47]. Adoption of supercritical CO_2 dyeing, a highly waterless and effluent-free dyeing process that combines low energy consumption and effective dye material recovery into the dye bath, prevents the generation of waste auxiliary chemicals other than waste dyes, and has the potential to be a novel sustainable process as long as scalability issues are resolved [11]. Together, these technologies, which include fiber modification, solvent enhancement, and advanced physical processing, are gradually gaining substantial ground in the direction of environmentally friendly textile dyeing and printing. High salt levels may be used in traditional reactive dyeing of cotton, which raises the TDS of wastewater. Pre-cationizing cotton fibers or using solvent-assisted dyeing media without the use of salt are examples of more modern, sustainable methods. In addition to providing effective dye exhaustion, these techniques significantly reduce water consumption, effluent pollution, and treatment costs while maintaining fabric quality, color fastness, and shade depth [44].

3.3.1 Natural Dyes with Bio-mordants

As people grow increasingly concerned about the health and environmental effects of employing metallic mordants and synthetic dyes, natural dyeing is gaining popularity. Other plant components that have bioactive chromophores (flavonoids, tannins, anthraquinones, and polyphenols) that can attach to fibers by hydrogen bonding, van der Waals forces, and ionic interactions include turmeric root (curcumin), date palm leaves, lotus seedpod waste, thyme, and catechu. However, mordants are necessary, and they are not very effective against cotton and its blends [16, 50]. Bio-mordants are renewable biopolymers and phytochemicals such as chitosan, tannins, and eucalyptus that can be used as environmentally friendly alternatives to conventional metallic mordants [6]. Chitosan, a cationic polysaccharide, develops strong hydrogen as well as covalent bonding with cellulosic fibers, which facilitates fixation and improves dye uptake due to an ionic mixed-polarity adsorption with anionic dye moieties. Tannins are natural polyphenolic compounds that chelate the dye-fiber complex as well as act as antibacterial and antioxidant compounds. Similarly, polyphenolic compounds contained in eucalyptus leaf extracts have two properties that they accelerate adhesion and produce biofunctional properties [19].

3.3.2 Cold Pad-Batch Reactive Dyeing

The best available technique for environmentally friendly cotton dyeing, according to industrial ecological concepts, is CPB reactive dyeing. Because it doesn't require dangerous auxiliaries or waste extra energy or water, and because a multifunctional element may be achieved without compromising the beauty of fabric, it satisfies many of the green chemistry pillars. In order to further improve resource efficiency and reduce batching time, and make the process more competitive in big volume manufacturing, recent research has also suggested combining CPB with enzymatic pretreatments or low-liquor application techniques [51]. Cold pad-batch (CPB) is a clean production technique that uses reactive dyes to color cellulosic fibers, particularly cotton. Here, the fabric is treated with an alkali solution and reactive dye at room temperature. It is then batched or stored in a plastic-covered roll to allow for fixing over a certain amount of time. CPB differs from traditional exhaust dyeing in that it does not require high temperatures or extra electrolytes, such as sodium chloride or sodium sulfate, which are prominent sources of total dissolved solids in textile effluent [57]. The fact that CPB is compatible with sustainable processing is among its biggest advantages. Since the dyeing process is carried out at a temperature that can be reached, no steam or heating energy is needed, which results in significant savings. When compared to exhaust dyeing, TCEO uses significantly less water due to its limited liquor ratio and lack of electrolytes. The environmental effect of textile wet processing can be reduced by using CPB dyeing, which has been shown to reduce effluent load and save up to 40% of water when compared to the original process [36].

3.3.3 Heritage Printing Methods: Dabu Printing

Dabu printing, a traditional Rajasthani hand-block resist printing technique, is well-known for being environmentally friendly. Using hand-carved wooden blocks, artisan groups also apply a resist paste made of mud (kali mitti), lime (chuna), natural gum, and wheat chaff. To aid in drying and provide additional adherence, sawdust is added to the paste. Because biodegradable and locally accessible materials are the only options and there is minimal exposure to industrial chemicals or energy-intensive manufacture, this soil-based approach is environmentally sustainable. Layers of color are produced using the resist and natural dyes, such as indigo, without the need for aggregate manufacturing [23]. In addition to being environmentally friendly, Dabu printing is cost-effective in that it uses sunshine for drying rather than mechanical heat and reuses dye vats, such as indigo, for extended periods of time with little lime or jaggery added. In keeping with the notion of circularity in the textile process, this will lessen the production of effluent [63].

3.4 Sustainable Finishing Treatments for Textiles

Finishing is a phase of textile production that requires a lot of resources and chemicals, making it a promising area for decarbonization, water sustainability, and toxic removal. Any sustainability plan should prioritize chemical replacement, process intensification, and closed-loop auxiliaries since wet-processing/finishing units regularly dominate life-cycle studies from gate to gate [66]. Dry or near-dry surface applications provide wettability and bonding sites in the absence of large water loads. They allow sustainable deposition of functional thin films within ambient conditions and reduce an auxiliary requirement as an instant control to minimize effluents [39].

3.4.1 Enzyme-Based Finishing

Cellulosic textile manufacturing, or more accurately, processing, has gained popularity. Enzyme-based finishing has emerged as a viable option for sustainable cellulosic textile finishing since it is environmentally friendly and more effective than conventional chemical processing. Cellulases are widely used in bio-polishing, which gives fabric a smooth, soft handle and a pleasing look by consistently and carefully breaking down surface fibrils [27]. In addition to adding antibacterial properties to textiles without the use of hazardous additives, laccases and peroxidases are increasingly being employed to graft functional

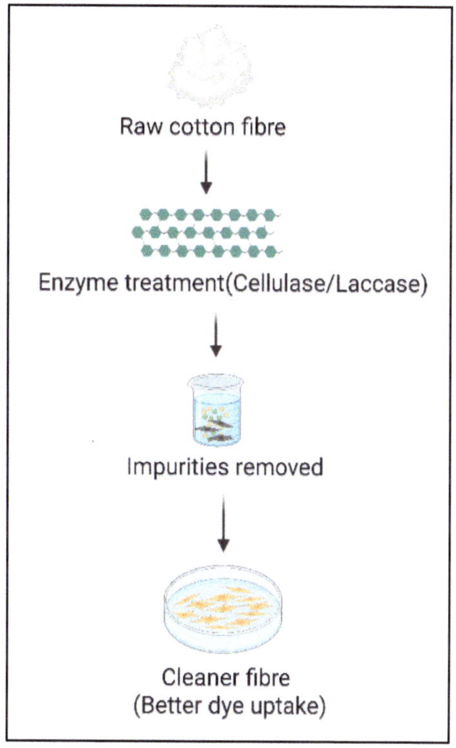

Fig. 3.2 Enzyme utilization in dyeing

groups. By breaking down any leftover hydrogen peroxide after bleaching, the catalases greatly aid in the elimination of peroxide and lessen the quantity of chemical required for the subsequent dyeing procedure. In addition to operating at softer temperatures and pH levels, these bioperatal processes can often improve the depth of the dye shade and hand feel while lowering the COD/BOD in effluents [14]. The enzymatic dying process is illustrated in Fig. 3.2.

3.4.2 Plasma Treatments

Reactive species (ions, radicals, and UV photons) can be generated by cold/low-temperature and atmospheric plasma, which are surface-only, dry processes that clean, etch, or functionalize textile surfaces without changing their bulk characteristics. Plasma finishing is promoted as a cleaner option than wet chemical baths since it requires less water and auxiliary materials and takes less time to complete [38]. The partially ionized gas known as low-temperature or cold plasma, which contains electrons, ions, radicals, and UV photons, can change the surface of textile fibers at the nanoscale without changing the material's bulk characteristics. Bombardment of reactive species to oxidize, cross-link, activate, etch, and polymerize the surface [5]. The energetic particles and reactive oxygen/nitrogen species found in plasma adsorb contaminants from the surface, remove micro-features, and provide polar functional groups that make surfaces more energetic and enable covalent grafting or improve wetting and dye penetration. The majority of mechanical and bulk properties are maintained, and all of these impacts are restricted to the uppermost layers [10]. Adding water/oil repellency and self-cleaning qualities to textile materials can be done environmentally with plasma finishing. By activating the fiber surface to introduce reactive sites that enable strong bonding through plasma polymerization or a little chemical addition, plasma-treated fibers guarantee good anchoring of thin repellent coatings. Without compromising handling of fabric, this increases liquid resistance. The idea of using plasma treatment as an environmentally benign substitute for fluorochemical-based finishing agents has been further validated by recent studies demonstrating that defensive textiles treated with plasma have higher self-cleaning qualities with less chemical application [1].

3.4.3 Advanced Oxidation Processes Using UV and Ozone

UV treatment often finds application in textiles, where surface functional groups are made to form reactive groups. Such reactive sites enhance fixing functional finishes, providing better wettability, dyeability, and coating adhesion. Importantly, this is a dry and chemical-free process, and does not impair the fabric strength and limits environmental load [34]. Ozone and UV radiation combine to generate hydroxyl radicals (·OH), some of the most potent oxidizers. Cotton and linen textiles can be bleached and surface-treated in an environmentally responsible manner because of the proven ability of radicals to effectively remove contaminants. According to a new study, UV/ozone bleaching systems can provide bleaching levels comparable to those of hydrogen peroxide bleaches without producing harmful effluent, which lessens their environmental impact [24]. UV/ozone finishing further enhances antimicrobial surfaces' effectiveness and longevity. For instance, even after extended exposure to UV and ozone, the ZnO sol-gel coatings on cotton textiles maintained their strong antibacterial activity against *Escherichia coli* and *Staphylococcus aureus*. This demonstrates that enhanced oxidation helps create self-cleaning surfaces as well as durable antimicrobial textiles [33]. The UV/ozone processes utilize much less water than wet processes and do not necessitate the use of hazardous chemical baths as in standard wet finishing operations. Ozone degrades spontaneously through an automated process that doesn't involve the addition of dangerous materials. UV/ozone and other more recent treatment techniques, when coupled with shorter processing times and less energy consumption, offer a sustainable and scalable substitute for textile finishing that adheres to green chemistry and circular economy concepts [34].

3.4.4 Nanotechnology-Based Green Finishes

The manufacture of green biogenic nanoparticles uses bacteria, fungi, algae, plants, or biomolecules that can act as reducing or capping agents and produce nanostructures based on ZnO, TiO_2, silver, and chitosan with minimal amounts of toxic chemicals. These nanoparticle (NP) finishes are some of the most widely used to enable multifunctional fabrics and lessen the total environmental impact of traditional chemical processes [12].

The green synthesis of ZnO NPs typically uses microbial metabolites or plant extracts, with riboflavin acting as a capping and reducing element to produce crystalline particles with a size of 10–100 nm. Techniques, including in-situ, dip-pad-dry-cure, and sol-gel deposition, and layer-by-layer assembly, are employed in textile film to increase toughness and adherence to fibers. ZnO NPs improve textile performance in an environmentally friendly way by providing antibacterial, UV, photocatalytic self-cleaning, and anti-biofilm qualities [15, 21].

3.5 Plasma and Nanotechnology Applications

3.5.1 Applications of Plasma Technology

Application of low temperature or oxygen plasma improves the wettability of cotton and other textile fibers considerably, and dyeability. Plasma treatment also confers polar functional groups on the fiber surface and raises the surface energy and hydrophilicity. Such an alteration can provide enhanced uptake and retention of dyes, contributing to the magnificent depth and consistency of most shades. The activation of plasma, therefore, minimizes the use of large liquor ratios, salt, and energy when dyeing, making that process more economical and sustainable towards the environment [25, 54].

Antimicrobial finishing based on plasma technology offers a more environmentally friendly alternative to traditional resin-based procedures. Silver (Ag) nanoparticles and other functional agents can be incorporated onto textile surfaces by the use of spark plasma treatment or plasma-assisted deposition. This method ensures strong surface adherence and long-lasting antibacterial activity without the use of hazardous binders or excessive chemicals. The resultant textiles maintain their softness and breathability while exhibiting long-lasting antibacterial and antifungal qualities. Thus, plasma-assisted antimicrobial finishing enhances the hygienic properties of textiles while also being environmentally friendly [38]. Both atmospheric and low-pressure plasma pretreatment work well to increase the adherence of coatings, inks, and printed electronics on textile materials. Through the introduction of reactive chemical functional groups and increased roughness, plasma is utilized to optimize the fiber's surface and improve bonding with applied layers. As a result, conductive inks, metallic pigments, and nanocoatings adhere better, improving electrical conductivity, durability, and wash-fastness. This change is particularly appealing for wearable electronics and smart fabrics, where long-term operation depends on coating adherence [13, 22].

By controlling the properties of the materials, atmospheric-pressure plasma (APP) treatments provide an eco-friendly method of preserving the longevity and aesthetic appeal of clothing. In textile wear and maintenance, APP eliminates recurring issues like pilling and hair falling by rearranging fiber cohesiveness and controlling surface roughness. Furthermore, it is possible to carefully refine plasma treatment to create the best possible fabric handle, drape, and stitchability without compromising comfort. In addition to minimizing fiber damage and ensuring long-lasting mechanical improvements, this chemical-free process contributes to the production of sustainable, higher-quality textiles with improved performance and consumer value [3, 10].

Advanced techniques like plasma polymerization and plasma-enhanced grafting are also employed to give textiles long-lasting functionality without affecting the bulk characteristics of the product. Reactive sites created by plasma activation can also be utilized to covalently graft or deposit polymer thin films, enabling fire-retardant, hydrophobic, stimuli-responsive, or hydrophilic properties. High resistance to washing and wear is the outcome of this method, which also offers even fiber coverage and exceptional bonding strength. In order to meet the sustainable intensive cleaning procedures of the textile industry, plasma-based grafting is more environmentally friendly, more energetically efficient, and does not require solvents like traditional wet-chemical finishing methods [48, 49]. The plasma pretreatment also promotes sustainable wet processing since it reduces the need for salts, auxiliaries, and intense scouring while increasing dye exhaustion. Shorter and colder dyeing procedures use less water and energy thanks to surface activation that improves wettability and dye uptake. Such a dry or nearly dry technique drastically cuts down on processing time, chemical use, and effluent amounts without compromising fiber quality. The capacity of plasma to replace or do away with conventional, economically consumptive wet procedures like scouring and desizing results in more inexpensive and ecologically friendly textile dyeing and finishing [65].

According to the latest research, optimizing the biodegradation of textile fibers, and reducing microplastic shedding may be facilitated by oxygen or air plasma treatment. By altering the surface's chemistry and topography, plasma enhances fiber cohesiveness and reduces micro-release during washing and wear. It can also be used to provide polar moieties to synthetic-natural blends, which will promote microbial contact and speed up biodegradation under favorable conditions. The future viability of textile materials and environmental degradation caused by microplastics are addressed by the combination of these two characteristics [28, 39].

3.5.2 Applications of Nanotechnology

Textiles are finished with copper nanoparticles, zinc oxide, and silver to provide long-lasting antibacterial protection. When it comes to protective masks, sportswear, and healthcare textiles, where odor and hygiene are crucial, such finishes can be a huge value addition. Application techniques include pad-dry-cure, in-situ nanoparticle processing on fibers, and binder or encapsulating procedures to increase wash fastness and prolong efficacy. These kinds of nanotechnology-based treatments provide high-performance, long-lasting antimicrobial protection without substantially altering the functionality or look of the fabric they are applied to [30, 52]. Nanoscale metal oxides like ZnO and TiO_2 that scatter and absorb UV light decisively have been used to achieve UV protection in textiles. These nanoparticles can raise an Ultraviolet Protection Factor of fabric without negatively affecting its breathability, comfort, or handiness. Durability, safety, and sustainability of the sunscreening performance are achieved using environmentally friendly synthesis techniques and surface treatment to improve particle-fiber compatibility [2].

Self-cleaning and photocatalytic coatings on textiles are made with TiO_2 nanoparticles and nanoparticle mixtures with other materials. These coatings break down organic debris, bacteria, and stains when exposed to UV or visible light. Because of this, they are ideal for technical textiles that require minimum maintenance, hospital furnishings, and outdoor materials. Improved photocatalytic activity and visible-light responsiveness can also be achieved by using carbon nanomaterials, like graphene or CNTs. All of these cutting-edge finishes are practical and eco-friendly textile features since they not only provide unparalleled cleaning economy and minimal chemical usage, but they also have antibacterial properties [58].

Nanoclays, layered silicates, nano-phosphates, and nanoscale intumescent systems are used to create multipurpose protective coatings and flame retardancy in textiles. The nanomaterials, which are deposited using sol-gel or layer-by-layer processes, are utilized to decrease flammability and enhance thermal stability. They may be made to have additional desirable qualities, such as water resistance, UV protection, or antibacterial activity, in addition to flame resistance. High-powered protective fabrics that support safety, durability, and sustainable performance are made available by this multifunctionality to the military, healthcare, and technology sectors [35, 56].

In terms of mechanical qualities, nanocellulose and nanofibers play a significant role in textile functionalization and reinforcement. As renewable, lightweight, and high-strength materials, they are used as coatings, additives, or sustainable bindings to give textiles strength, abrasion resistance, wrinkle resistance, and general handle. Their nanoscale size allows them to print easily on surfaces and improves breathability and smoothness. Additionally, nanocellulose can serve as an expanding carrier framework for additional nanofunctional additives, enabling the development of multifunctional qualities like flame retardancy, UV protection, and antibacterial activity into an environmentally benign, eco-diversified textile [32, 59].

Core-shell polymeric nanoparticles, hydrophobic silica nanoparticles, and nanoroughness surface coatings are becoming popular methods for giving textiles stain and water resistance. These techniques are frequently used in conjunction with fluorine-free chemistries to create durable protective barriers that limit environmental damage and withstand liquid penetration. Long-term adherence and performance are enhanced by the combination of plasma pretreatment and nanoparticle deposition. In addition to using fewer chemicals than traditional treatments, these finishes also focus on maintaining comfort, breathability, and appearance of fabric, making it suitable for outdoor wear, upholstery, and the use of technical protective textiles [60, 61].

Nano-encapsulation into polymers or inorganic carriers enables the application of controlled release on textiles; this method enables the incorporation of phase-change materials, perfumes, antimicrobials, and UV absorbers. This procedure offers long-lasting, prolonged life and qualities that are resistant to washing. Encapsulated systems provide long-lasting comfort, freshness, and temperature management to a broader range of applications, such as sportswear, protective apparel, and health care, particularly in smart and functional textiles [4].

3.6 Conclusion

Sustainable chemical processing is a recent development in textile processing that introduces the idea of responsible performance. Nanotechnology, plasma, UV/ozone finishing, supercritical CO_2 dyeing, and enzyme pretreatments enable energy and water-efficient processes with less hazardous byproducts. The procedures replace hazardous substances with more comfortable, adaptable, and long-lasting alternatives. Bio-mordanted dyes, bio-based repellents (non-PFAS repellents), and bio-based cross-linkers are examples of renewable sources that are being used in the mainstream. Similar to this, a revival of traditional designs like Dabu printing highlights the benefits of sustainable craft for the environment and culture in a modern setting. Even though there are still issues with the scalability of dyeing, cost, and compatibility, particularly with more sophisticated techniques like supercritical dyeing or plasma-based dyeing, increasing research and industrial testing are offering more potential causal proxies. In the end, switching to eco-friendly dyes or textile finishing not only becomes necessary for the environment but also offers an opportunity to reconsider the market's competitiveness, which is increasingly centered on sustainability factors. The textile sector will be able to meet consumer demands and the agenda for global sustainable growth by utilizing green chemistry concepts and technology advancements to provide safer products, a more circular economy, and a smaller environmental impact.

References

1. Akter, S., Mahmud, Md. F., Rahman, A. N. M. M., Pritha, N. M., Hasan, Md. M., Ullah, Md. H., Kanon, Md. R., Ahona, F. T., & Bristy, B. F. (2024). Application of Low-Temperature Air Plasma for the Enhancement of Defense Fabric's Self-Cleaning Property. *ACS Omega, 9*(39), 41053–41066. https://doi.org/10.1021/acsomega.4c06893
2. Asmat-Campos, D., Delfín-Narciso, D., & Juárez-Cortijo, L. (2021). Textiles Functionalized with ZnO Nanoparticles Obtained by Chemical and Green Synthesis Protocols: Evaluation of the Type of Textile and Resistance to UV Radiation. *Fibers, 9*(2), 10. https://doi.org/10.3390/fib9020010
3. Barni, R., Roman, H. E., Citterio, A., Leonardi, G., & Riccardi, C. (2022). Atmospheric plasma treatments of cashmere: The role of nanoscale sizing in the spray coating processing. *Frontiers in Materials, 9.* https://doi.org/10.3389/fmats.2022.987608
4. Bárta, J., Hájková, K., Sikora, A., Jurczyková, T., Popelková, D., & Kalous, P. (2023). Effect of a Nanocellulose Addition on the Mechanical Properties of Paper. *Polymers, 16*(1), 73. https://doi.org/10.3390/polym16010073
5. Bisoyi, D. K. (2023). A Review of Cold Plasma Treatment on Textile Fibre. *Trends in Textile Engineering & Fashion Technology, 7*(5). https://doi.org/10.31031/TTEFT.2023.07.000675
6. Brudzyńska, P., Sionkowska, A., & Grisel, M. (2023). Cotton Textile Dyeing by Plant-Derived Colorants in the Presence of Natural Additives. *Fibers and Polymers, 24*(10), 3641–3655. https://doi.org/10.1007/s12221-023-00332-3
7. Castillo-Suárez, L. A., Sierra-Sánchez, A. G., Linares-Hernández, I., Martínez-Miranda, V., & Teutli-Sequeira, E. A. (2023). A critical review of textile industry wastewater: Green technologies for the removal of indigo dyes. *International Journal of Environmental Science and Technology, 20*(9), 10553–10590. https://doi.org/10.1007/s13762-023-04810-2
8. Catarino, M. L., Sampaio, F., & Gonçalves, A. L. (2025). Sustainable Wet Processing Technologies for the Textile Industry: A Comprehensive Review. *Sustainability, 17*(7), 3041. https://doi.org/10.3390/su17073041
9. Chauhan, V. K., Dubey, R., Pandey, R., Thennarasu, P., & Dubey, S. K. (2024). MXene on Textile Substrates: A Review. *Advanced Engineering Materials, 26*(7). https://doi.org/10.1002/adem.202301962
10. Chen, Z., Zhao, Z., & Wu, J. (2023). Application of Atmospheric-Pressure Plasma Treatment in Anti-Hairfalling of Polyester–Cotton Fleece Knitted Fabrics. *Polymers, 15*(9), 2097. https://doi.org/10.3390/polym15092097
11. de Oliveira, C. R. S., de Oliveira, P. V., Pellenz, L., de Aguiar, C. R. L., & da Silva Júnior, A. H. (2024). Supercritical fluid technology as a sustainable alternative method for textile dyeing: An approach on waste, energy, and CO2 emission reduction. *Journal of Environmental Sciences, 140*, 123–145. https://doi.org/10.1016/j.jes.2023.06.007
12. Dejene, B. K. (2024). Biosynthesized ZnO nanoparticle-functionalized fabrics for antibacterial and biocompatibility evaluations in medical applications: A critical review. *Materials Today Chemistry, 42*, 102421. https://doi.org/10.1016/j.mtchem.2024.102421
13. Donkor, S. O., Sun, D., Bucknall, D., & Buckman, J. (2024). Surface Activation of Cotton Fabric with Low-Temperature Air Plasma Treatment for Metallic Printing. *Fibers and Polymers, 25*(11), 4321–4332. https://doi.org/10.1007/s12221-024-00738-7
14. Eid, B. M., & Ibrahim, N. A. (2021). Recent developments in sustainable finishing of cellulosic textiles employing biotechnology. *Journal of Cleaner Production, 284*, 124701. https://doi.org/10.1016/j.jclepro.2020.124701

15. El-Fallal, A. A., Elfayoumy, R. A., & El-Zahed, M. M. (2023). Antibacterial activity of biosynthesized zinc oxide nanoparticles using Kombucha extract. *SN Applied Sciences*, *5*(12), 332. https://doi.org/10.1007/s42452-023-05546-x
16. Fang, J., Meng, C., Wang, Y., Yang, Y., Han, L., Wang, S., Zhang, G., Xu, Z., & Min, J. (2023). Eco-dyeing and Functional Finishing of Cotton Fabric by Natural Dye Derived from Lotus Seedpod Waste with Chitosan-Assistance. *Fibers and Polymers*, *24*(4), 1367–1377. https://doi.org/10.1007/s12221-023-00158-z
17. Fu, Z., Sun, K., Fan, H., Li, C., Liu, H., Zhang, S., Ding, K., Gao, G., & Hu, X. (2022). Understanding evolution of the products and emissions during chemical activation of furfural residue with varied potassium salts. *Journal of Cleaner Production*, *357*, 131936. https://doi.org/10.1016/j.jclepro.2022.131936
18. Ge, H., Liu, Y., & Liu, F. (2023). Up to Date Review of Nature-Inspired Superhydrophobic Textiles: Fabrication and Applications. *Materials*, *16*(21), 7015. https://doi.org/10.3390/ma16217015
19. Ghosh, J., Khan, F., Noor, T., Rupanty, N. S., & Das, S. C. (2024). Union Dyeing of Cotton-Silk Blended Woven Fabric Using Tanin Bio-mordant and Thyme as a Colourant. *Textile & Leather Review*, *7*, 303–326. https://doi.org/10.31881/TLR.2023.208
20. Ghosh, J., Rupanty, N. S., Khan, F., Noor, T., Jahangir, R., Mirmohammadsadeghi, S., & Islam, T. (2025). Grafting modification for textile functionalization: Innovations and applications. *Discover Applied Sciences*, *7*(1), 49. https://doi.org/10.1007/s42452-024-06364-5
21. Giedraitienė, A., Ružauskas, M., Šiugždinienė, R., Tučkutė, S., Grigonis, K., & Milčius, D. (2024). ZnO Nanoparticles Enhance the Antimicrobial Properties of Two-Sided-Coated Cotton Textile. *Nanomaterials*, *14*(15), 1264. https://doi.org/10.3390/nano14151264
22. Guérineau, J., Ton, J., & Zhuldybina, M. (2025). Screen Printing Conductive Inks on Textiles: Impact of Plasma Treatment. *Sensors*, *25*(13), 4240. https://doi.org/10.3390/s25134240
23. Hada, J. S., & Meena, C. R. (2022). *Dabu, The Sustainable Resist Printed Fabric of Rajasthan* (pp. 69–91). https://doi.org/10.1007/978-981-19-0538-4_3
24. Hamada, K., Ochiai, T., Tsuchida, Y., Miyano, K., Ishikawa, Y., Nagura, T., & Kimura, N. (2020). Eco-Friendly Cotton/Linen Fabric Treatment Using Aqueous Ozone and Ultraviolet Photolysis. *Catalysts*, *10*(11), 1265. https://doi.org/10.3390/catal10111265
25. Haque, A. N. M. A., & Naebe, M. (2023). Zero-water discharge and rapid natural dyeing of wool by plasma-assisted spray-dyeing. *Journal of Cleaner Production*, *402*, 136807. https://doi.org/10.1016/j.jclepro.2023.136807
26. Hashem, A., & Farag, S. (2025). Cotton bleaching: Evolution, current practices, and future perspectives. *The International Journal of Advanced Manufacturing Technology*. https://doi.org/10.1007/s00170-025-16185-4
27. Ikbal, Md. S., Tisha, F. A., Asheque, A. I., Hasnat, E., & Uddin, M. A. (2024). Eco-friendly biopolishing of cotton fabric through wasted sugarcane bagasse-derived enzymes. *Heliyon*, *10*(4), e26346. https://doi.org/10.1016/j.heliyon.2024.e26346
28. Jabbar, A., Bryant, M., Armitage, J., & Tausif, M. (2024). Oxygen plasma treatment to mitigate the shedding of fragmented fibres (microplastics) from polyester textiles. *Cleaner Engineering and Technology*, *23*, 100851. https://doi.org/10.1016/j.clet.2024.100851
29. Jabbar, M., Iftikhar, F., Hussain, T., Ahmad, S., & Ahmad, F. (2025). Preparation and characterization of sustainable plant-based blended woven fabric. *Journal of Engineered Fibers and Fabrics*, *20*. https://doi.org/10.1177/15589250251319322
30. Jain, A., Kongkham, B., Puttaswamy, H., Butola, B. S., Malik, H. K., & Malik, A. (2022). Development of Wash-Durable Antimicrobial Cotton Fabrics by In Situ Green Synthesis of Silver Nanoparticles and Investigation of Their Antimicrobial Efficacy against Drug-Resistant Bacteria. *Antibiotics*, *11*(7), 864. https://doi.org/10.3390/antibiotics11070864

31. Jin, L., Ji, C., Chen, S., Song, Z., Zhou, J., Qian, K., & Guo, W. (2023). Multifunctional Textiles with Flame Retardant and Antibacterial Properties: A Review. *Molecules*, *28*(18), 6628. https://doi.org/10.3390/molecules28186628
32. Khalid, M. Y., Al Rashid, A., Arif, Z. U., Ahmed, W., & Arshad, H. (2021). Recent advances in nanocellulose-based different biomaterials: Types, properties, and emerging applications. *Journal of Materials Research and Technology*, *14*, 2601–2623. https://doi.org/10.1016/j.jmrt.2021.07.128
33. Kim, J.-H., Ma, J., Lee, S., Jo, S., & Kim, C. S. (2019). Effect of Ultraviolet–Ozone Treatment on the Properties and Antibacterial Activity of Zinc Oxide Sol-Gel Film. *Materials*, *12*(15), 2422. https://doi.org/10.3390/ma12152422
34. Korkmaz, G., Kılınç, M., Kılınç, N., & Dilek Kut, Y. (2024). The Role of Surface Modification Methods for Sustainable Textiles. In *Roadmap to Sustainable Textiles*. IntechOpen. https://doi.org/10.5772/intechopen.112792
35. Magovac, E., Vončina, B., Jordanov, I., Grunlan, J. C., & Bischof, S. (2022). Layer-by-Layer Deposition: A Promising Environmentally Benign Flame-Retardant Treatment for Cotton, Polyester, Polyamide and Blended Textiles, *Materials*, *15*(2), 432. https://doi.org/10.3390/ma15020432
36. Majeed, A., Carmenini, R., Locatelli, E., Sambri, L., & Comes Franchini, M. (2025). Salt-free dyeing of cotton fabric via graft polymerization with chitosan using dimethyl itaconate as a renewable cross-linker. *RSC Advances*, *15*(31), 25733–25741. https://doi.org/10.1039/D5RA02510A
37. Monisha, A. S., Rajeswari, S., Parasakthibala, G., & Dhanalakshmi, P. (2023). A View on Recent Finishes in Textiles. *International Journal for Research in Applied Science and Engineering Technology*, *11*(9), 410–413. https://doi.org/10.22214/ijraset.2023.55670
38. Naebe, M., Haque, A. N. M. A., & Haji, A. (2022). Plasma-Assisted Antimicrobial Finishing of Textiles: A Review. *Engineering*, *12*, 145–163. https://doi.org/10.1016/j.eng.2021.01.011
39. Nyssanbek, M., Kuzina, N., Kondrashchenko, V., & Azimov, A. (2024). Effects of plasma treatment on biodegradation of natural and synthetic fibers. *Npj Materials Degradation*, *8*(1), 23. https://doi.org/10.1038/s41529-024-00437-x
40. Orasugh, J. T., Temane, L. T., Kesavan Pillai, S., & Ray, S. S. (2025). Advancements in Antimicrobial Textiles: Fabrication, Mechanisms of Action, and Applications. *ACS Omega*, *10*(13), 12772–12816. https://doi.org/10.1021/acsomega.4c11356
41. Pan, Y., Fu, L., Du, J., Zhang, D., Lü, T., Zhang, Y., & Zhao, H. (2022). Layer-by-Layer Self-Assembly Coating for Multi-Functionalized Fabrics: A Scientometric Analysis in CiteSpace (2005–2021). *Molecules*, *27*(19), 6767. https://doi.org/10.3390/molecules27196767
42. Pandit, P., Singha, K., & Maity, S. (2021). Green chemistry in textile and fashion. In *Chemical Management in Textiles and Fashion* (pp. 177–203). Elsevier. https://doi.org/10.1016/B978-0-12-820494-8.00009-5
43. Patel, D. K., Ganguly, K., Hexiu, J., Dutta, S. D., Patil, T. V., & Lim, K.-T. (2022). Functionalized chitosan/spherical nanocellulose-based hydrogel with superior antibacterial efficiency for wound healing. *Carbohydrate Polymers*, *284*, 119202. https://doi.org/10.1016/j.carbpol.2022.119202
44. Patil, S., & Athalye, Dr. A. (2024). Sustainable Coloration Technologies. *Indian Journal of Fibre and Textile Engineering*, *4*(2), 1–8. https://doi.org/10.54105/ijfte.A2413.04021124
45. Patti, A. (2025). Green Advances in Wet Finishing Methods and Nanoparticles for Daily Textiles. *Macromolecular Rapid Communications*, *46*(2). https://doi.org/10.1002/marc.202400636
46. Periyasamy, A. P. (2024). Recent Advances in the Remediation of Textile-Dye-Containing Wastewater: Prioritizing Human Health and Sustainable Wastewater Treatment. *Sustainability*, *16*(2), 495. https://doi.org/10.3390/su16020495

47. Periyasamy, A. P., Dutta, S., & Tehrani-Bagha, A. R. (2024). Solvent-assisted salt-free reactive dyeing of cotton fabric. *Cellulose, 31*(11), 7101–7118. https://doi.org/10.1007/s10570-024-05928-3
48. Primc, G., Zaplotnik, R., Vesel, A., & Mozetič, M. (2022). Mechanisms Involved in the Modification of Textiles by Non-Equilibrium Plasma Treatment. *Molecules, 27*(24), 9064. https://doi.org/10.3390/molecules27249064
49. Radetić, M., & Marković, D. (2022). A review on the role of plasma technology in the nano-finishing of textile materials with metal and metal oxide nanoparticles. *Plasma Processes and Polymers, 19*(4). https://doi.org/10.1002/ppap.202100197
50. Rahman, M. M., Kim, M., Youm, K., Kumar, S., Koh, J., & Hong, K. H. (2023). Sustainable one-bath natural dyeing of cotton fabric using turmeric root extract and chitosan biomordant. *Journal of Cleaner Production, 382*, 135303. https://doi.org/10.1016/j.jclepro.2022.135303
51. Ristić, I., Zdravković, A., Mičić, A., Marković, N. D., & Ristić, N. (2020). Ecological alternatives to conventional dyeing of cotton with reactive dyes. *Zastita Materijala, 61*(1), 60–68. https://doi.org/10.5937/zasmat2001060R
52. Rodrigues, A. S., Batista, J. G. S., Rodrigues, M. Á. V., Thipe, V. C., Minarini, L. A. R., Lopes, P. S., & Lugão, A. B. (2024). Advances in silver nanoparticles: A comprehensive review on their potential as antimicrobial agents and their mechanisms of action elucidated by proteomics. *Frontiers in Microbiology, 15*. https://doi.org/10.3389/fmicb.2024.1440065
53. Rungruangkitkrai, N., Phromphen, P., Chartvivatpornchai, N., Srisa, A., Laorenza, Y., Wongphan, P., & Harnkarnsujarit, N. (2024). Water Repellent Coating in Textile, Paper and Bioplastic Polymers: A Comprehensive Review. *Polymers, 16*(19), 2790. https://doi.org/10.3390/polym16192790
54. Serrano-Martínez, V. M., Ruzafa-Silvestre, C., Hernández-Fernández, C., Bañón-Gil, E., Arán-Ais, F., & Orgilés-Calpena, E. (2024). Improving Cotton Fabric Dyeability by Oxygen Plasma Surface Activation. *Surfaces, 7*(4), 1079–1095. https://doi.org/10.3390/surfaces7040071
55. Sharif, R., Qutab, H. G., Mahmood, K., Gul, S., Ramzan, N., Mohsin, M., Wahlah, A., Nasir, R., Fazal, P., & Ali, B. (2024). One pot application of a green chemistry-based finish for cotton fabric, providing hydrophobic, flame retardant, and antimicrobial properties. *RSC Advances, 14*(9), 6146–6155. https://doi.org/10.1039/D3RA07931G
56. Sharma, V., Agarwal, S., Mathur, A., Singhal, S., & Wadhwa, S. (2024). Advancements in nanomaterial based flame-retardants for polymers: A comprehensive overview. *Journal of Industrial and Engineering Chemistry, 133*, 38–52. https://doi.org/10.1016/j.jiec.2023.12.010
57. Shu, D., Fang, K., Liu, X., Cai, Y., Zhang, X., & Zhang, J. (2018). Cleaner coloration of cotton fabric with reactive dyes using a pad-batch-steam dyeing process. *Journal of Cleaner Production, 196*, 935–942. https://doi.org/10.1016/j.jclepro.2018.06.080
58. Solovyeva, M., Selishchev, D., Cherepanova, S., Stepanov, G., Zhuravlev, E., Richter, V., & Kozlov, D. (2020). Self-cleaning photoactive cotton fabric modified with nanocrystalline TiO2 for efficient degradation of volatile organic compounds and DNA contaminants. *Chemical Engineering Journal, 388*, 124167. https://doi.org/10.1016/j.cej.2020.124167
59. Spagnuolo, L., D'Orsi, R., & Operamolla, A. (2022). Nanocellulose for Paper and Textile Coating: The Importance of Surface Chemistry. *ChemPlusChem, 87*(8). https://doi.org/10.1002/cplu.202200204
60. Syduzzaman, Md., Hassan, A., Anik, H. R., Akter, M., & Islam, Md. R. (2023). Nanotechnology for High-Performance Textiles: A Promising Frontier for Innovation. *ChemNanoMat, 9*(9). https://doi.org/10.1002/cnma.202300205
61. Tamulevičius, T. (2022). Nanotechnologies in Textiles. *Materials, 15*(4), 1466. https://doi.org/10.3390/ma15041466

62. Taurino, R., Cannio, M., Martinuzzi, S., Caporali, S., Borgioli, F., Galvanetto, E., & Boccaccini, D. (2025). Enhancing textile properties using nanotechnology and 3D printing process. *Journal of Sol-Gel Science and Technology*, *114*(3), 916–933. https://doi.org/10.1007/s10971-025-06749-x
63. Vankar, P. S., & Shukla, D. (2024). Traditional block printing for sustainability. In *Natural Dyes for Sustainable Textiles* (pp. 151–165). Elsevier. https://doi.org/10.1016/B978-0-323-85257-9.00003-7
64. Yacoub, A. A., & Patel, B. H. (2024). Atmospheric plasma technology for sustainable wet processing of textiles. *Materials Today: Proceedings.* https://doi.org/10.1016/j.matpr.2024.04.043
65. Zaidy, S. S. H., Vacchi, F. I., Umbuzeiro, G. A., & Freeman, H. S. (2019). Approach to Waterless Dyeing of Textile Substrates—Use of Atmospheric Plasma. *Industrial & Engineering Chemistry Research*, *58*(40), 18478–18487. https://doi.org/10.1021/acs.iecr.9b01260
66. Zhang, S., Xu, C., Xie, R., Yu, H., Sun, M., & Li, F. (2023). Environmental assessment of fabric wet processing from gate-to-gate perspective: Comparative study of weaving and materials. *Science of The Total Environment*, *857*, 159495. https://doi.org/10.1016/j.scitotenv.2022.159495
67. Zhou, X., Yu, Q., Li, Y., Yuan, Z., He, W., Zhang, L., & Cheng, Z. (2024). Durable superhydrophobic cotton fabrics based on fluorinated alternating copolymer and silica nanoparticle: Preparation and oil-water separation. *European Polymer Journal*, *211*, 113029. https://doi.org/10.1016/j.eurpolymj.2024.113029

Future Directions and Emerging Trends in Eco-Friendly Textiles

4

Contents

A. H. Bari et al., *Eco-Friendly Textile Processes*, Synthesis Lectures on Chemical Engineering and Biochemical Engineering, https://doi.org/10.1007/978-3-032-08916-8_4

Abstract

The evolution of textiles is increasingly driven by the need to balance advanced functionality with environmental responsibility. Smart and eco-friendly textiles combine responsive features such as thermal regulation, moisture control, and electronic conductivity with sustainability principles. These materials are designed using approaches like green chemistry, renewable and bio-derived fibers, and energy-efficient processing methods while addressing recyclability and end-of-life management. By integrating digital tools such as IoT, data analytics, and automated manufacturing, these textiles hold promise for applications in healthcare, protective clothing, sportswear, fashion, and environmental monitoring. Their development signals a pathway toward a more sustainable textile industry.

Keywords

Smart textiles • Eco-friendly fabrics • Responsive materials • Sustainable design • Conductive fibers • Thermal regulation • Renewable polymers • Circular economy • Wearable technology • Green processing

4.1 Introduction

The textile sector is steadily transitioning from conventional fabrics to advanced materials that integrate sustainability with multifunctionality. While traditional fabrics have primarily been limited to passive roles, such as providing clothing, protection, or aesthetic value, modern smart textiles embody a new generation of innovation. These textiles move beyond simple functionality by combining insights from materials science, nanotechnology, digital technologies, and eco-friendly chemistry. They possess the ability to sense external stimuli, including temperature, light, humidity, or pH, and can actively respond or adapt to changing conditions. This adaptability enhances wearer comfort, extends product usability, and supports broader goals of environmental responsibility [58, 62].

Two key drivers have largely influenced the advancement of smart textiles. First, there is an increasing need for high-performance apparel in specialized sectors such as healthcare, defense, and athletic wear, where functionality and adaptability are critical. Second, the global push for sustainability has made eco-innovation a central focus of textile research. In response, efforts are now directed toward incorporating renewable fibers, biodegradable polymers, and environmentally benign processing technologies into smart textile design. This dual emphasis enhances product performance and aligns the industry with long-term ecological and societal goals [23].

Smart and responsive textiles, when developed with eco-friendly principles, go beyond providing luxury or advanced functionality. Their design philosophy emphasizes reducing environmental impacts throughout the life cycle spanning raw material sourcing, manufacturing, usage, and disposal. This is achieved by incorporating biodegradable sensing elements, bio-derived nanomaterials, energy-efficient fabrication methods, and strategies for recyclability. By enabling properties such as self-cleaning, thermal regulation, moisture adaptability, and material recovery, these textiles present a transformative pathway toward sustainable innovation in the fabric industry and foster a material culture that balances performance with ecological responsibility [17, 60].

This chapter provides a comprehensive overview of the scientific foundations, materials, and technological innovations that define smart and responsive textiles. It examines their classification, fabrication techniques, and the integration of green chemistry approaches that ensure environmental compatibility. Emphasis is placed on eco-friendly pathways, including the use of renewable resources, sustainable processing, and circular economy strategies. The chapter also highlights diverse applications across healthcare, defense, sports, and environmental sectors, while evaluating their environmental impact through life cycle assessment. Finally, it outlines key challenges and future directions, offering insights into the development of sustainable, high-performance textile systems.

4.2 Classification of Smart and Responsive Textiles

Smart textiles can be broadly categorized into three generations:

1. **Passive Smart Textiles**

   The passive smart textile is a fabric designed to sense the external atmospheric conditions or stimuli. They protect against environment conditions such as temperature, microbes, ultraviolet (UV) radiation, and moisture in sweat. Interestingly, the passive smart textile does not depend on external power sources like electricity, sensors, or actuators [70]. Table 4.1 summarizes the different features of smart and responsive textiles.

2. **Active Smart Textiles**

   The active smart textiles can sense the external stimuli, such as temperature, microbes, ultraviolet (UV) radiation, and moisture, and immediately respond to the changes of such stimuli. Conversely to the passive textile, this textile needs external power sources like electricity, sensors, or actuators. For instance, thermochromic fabrics can provide visual changes by changing color in response to temperature variations. Integrating phase-change materials (PCM) into textiles can enhance comfort by adapting thermal insulation to environmental conditions [68].
3. **Ultra-Smart Textiles**

   The ultra-smart textiles are the most advanced class of active smart textiles, which can sense, respond, and adapt intelligently to stimuli. Such fabrics often integrate artificial intelligence (AI), enabling self-learning and predictive responses. Wearable biosensors for health monitoring are prime examples [5, 10].

From an eco-friendly standpoint, these classifications incorporate sustainability principles. For instance, thermochromic dyes derived from plant-based sources and biodegradable electronic circuits exemplify "green" smart textiles [13].

4.3 Material Innovations in Eco-Friendly Smart Textiles

4.3.1 Natural Fibers and Biopolymers

Scientists around the globe are focusing on protecting the atmosphere and biodiversity by enhancing the sustainability of eco-friendly products. Due to the biorenewable properties of natural fibers, it is paramount to replace synthetic materials with natural fibers. Although natural fibers have drawbacks like biocompatibility and hydrophilic properties, surface modifications and chemical treatments can improve these properties. Natural fibers are successfully used in various applications, including composite materials, textiles, and smart packaging. Their utilization can also help reduce pollution related to waste, landfills, and greenhouse gas emissions. Natural fibers such as cotton, hemp, jute, and bamboo remain foundational in eco-friendly smart textiles due to their biodegradability and renewability [28, 33]. Advances in surface functionalization enable these fibers to act as substrates for sensors or responsive coatings. Biopolymers such as polylactic acid (PLA) and chitosan are increasingly explored for embedding smart functionalities [38, 44].

Table 4.1 Different features of smart and responsive textiles

S. No.	Features	Smart and responsive textiles		
		Passive textiles	Smart textiles	Ultra-smart textiles
1	Functionality	Provides basic comfort, protection, or aestheticity	Responds to external stimuli (temperature, moisture, light, etc.)	Actively senses, processes, and adapts to stimuli in real time
2	Types of response	No response from the textile	One-way	Two-way, autonomous, or adaptive responses
3	Integration of technology	None	Embedded sensors, conductive fibers, or coatings	Advanced sensors, IoT, AI, actuators, energy harvesting, data processing
4	Requirement of energy	None	Low (minimal energy for actuation)	High (requires power or energy-harvesting modules)
5	Areas of application	Everyday clothing, basic protective textiles	Sportswear, healthcare textiles, adaptive clothing	Military, biomedical devices, smart healthcare systems, interactive fashion
6	Sustainability potential	Moderate (depends on material choice)	Can be eco-friendly if renewable or biodegradable materials used	High if combined with green fabrication, renewable materials, and IoT-driven efficiency
7	Cost and complexity	Low	Moderate	High
8	Examples of polymers/textiles	Cotton, wool, polyester	Phase-change materials, moisture-responsive fabrics	IoT-enabled garments, wearable health monitors, adaptive camouflage textiles

4.3.2 Nanomaterials

The properties of nanomaterials, including their durability, rapid response to external stimuli, and environmental sustainability, meet the growing demands of the textile industry. They are incorporated into textile products to enhance properties such as stain and moisture resistance, UV protection, antibacterial effects, and temperature regulation [55]. The following nanomaterials are used in the textile industry to improve their response to external stimuli.

- **Metal and Metal Oxide Nanoparticles**: Metal and metal oxide nanoparticles have revolutionized textile engineering by imparting antimicrobial, UV-protective, and self-cleaning properties. Silver nanoparticles exhibit potent antibacterial activity via ion release and oxidative stress, making them popular in medical and sports fabrics [66]. Copper and ZnO nanoparticles offer cost-effective alternatives with notable antimicrobial and UV-shielding potential [48, 65] TiO_2 nanoparticles provide photocatalytic self-cleaning and enhanced ultraviolet protection factors [30]. Advanced techniques such as sol-gel processing and in situ synthesis improve nanoparticle stability on fibers while reducing leaching [18]. Sustainable approaches using green synthesis and biopolymer stabilizers enhance safety and environmental compatibility [42].
- **Graphene and graphene oxide** Graphene and graphene oxide (GO) have gained significant attention in textile applications due to their unique mechanical, thermal, and electrical properties. Their incorporation into fabrics enhances strength, conductivity, and barrier functions while maintaining flexibility. Graphene-based coatings provide superior ultraviolet (UV) shielding and antibacterial activity, making them suitable for protective clothing and healthcare textiles. With abundant oxygen functional groups, GO offers excellent dispersibility and compatibility with polymers, enabling durable surface functionalization. Moreover, their integration supports smart textiles for wearable electronics, sensors, and energy storage. Sustainable synthesis and eco-friendly processing remain central to advancing graphene-based textiles toward commercial viability [22].
- **Cellulose nanofibers:** Cellulose nanofibers (CNFs) are renewable, lightweight, and high-strength materials derived from natural cellulose sources that are increasingly applied in textile engineering. Their nanoscale dimensions and abundant hydroxyl groups enable strong interfacial bonding with fibers, improving mechanical stability, durability, and moisture management. CNFs can be integrated into fabrics to impart biodegradability, enhanced barrier properties, and surface smoothness without compromising comfort. Additionally, CNFs serve as carriers for functional agents such as antimicrobial compounds or dyes, extending the performance of eco-friendly textiles. Their compatibility with green processing makes them highly attractive for sustainable apparel, technical textiles, and next-generation composite applications [25].

  Efforts are being made to replace heavy-metal nanoparticles with bio-derived alternatives such as lignin nanoparticles or plant-based quantum dots [47]. The need for nanotechnology in smart textiles is illustrated in Fig. 4.1.

4.3.3 Phase-Change Materials (PCMs)

Phase-change materials (PCMs) are widely explored in textiles for their ability to regulate temperature through reversible solid-liquid transitions. When integrated into

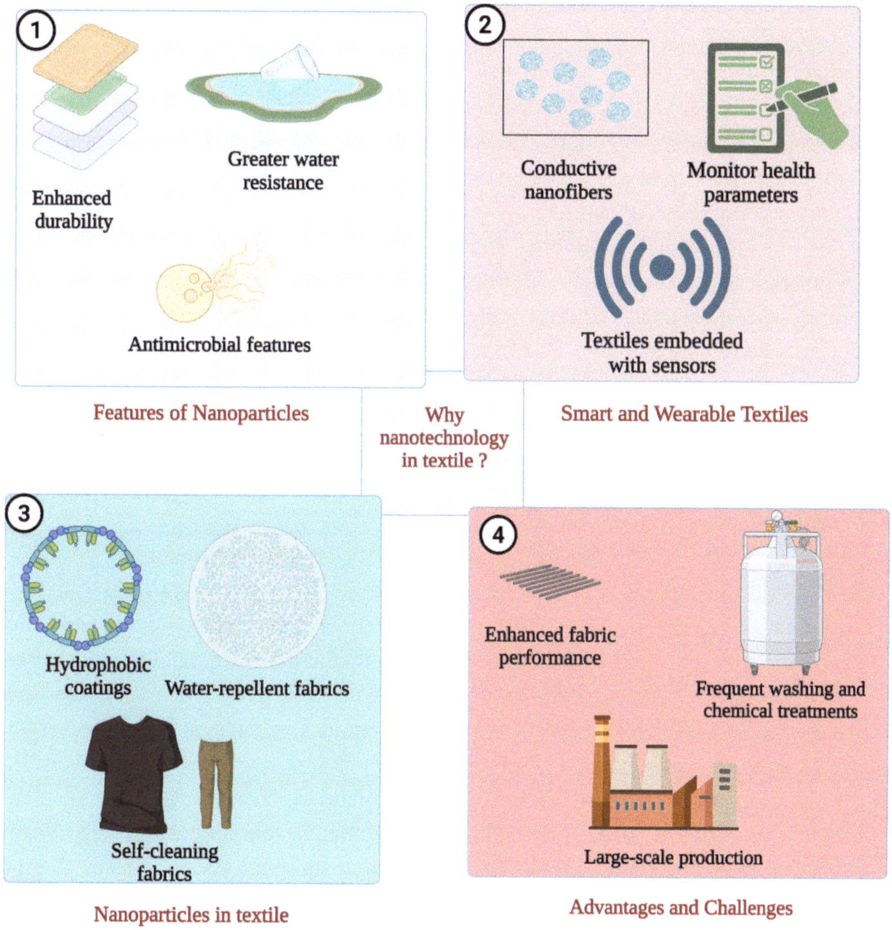

Fig. 4.1 Need and role of nanotechnology in smart textiles

fibers, coatings, or microcapsules, PCMs absorb, store, and release latent heat, enhancing wearer comfort under fluctuating environmental conditions. Commonly used PCMs include paraffins, fatty acids, and hydrated salts, which can be incorporated into fabrics using microencapsulation or impregnation techniques. Advances in bio-based PCMs and nanostructured carriers improve thermal stability, durability, and wash resistance. PCM-functionalized textiles hold promise for sportswear, protective clothing, and smart fabrics, offering adaptive thermal regulation while aligning with sustainable design principles [63, 67].

4.3.4 Conductive Polymers

Conductive polymers (CPs) such as polyaniline, polypyrrole, and poly(3,4-ethylenedioxythiophene) (PEDOT) have emerged as functional materials for smart textiles due to their intrinsic electrical conductivity and flexibility. When deposited onto fabrics by in situ polymerization, coating, or printing, CPs impart electroactivity without significantly altering textile comfort. These materials enable wearable sensors, energy storage, and electromagnetic shielding applications. Unlike metallic coatings, CPs offer lightweight and corrosion-resistant alternatives. Recent research focuses on improving wash durability, mechanical stability, and eco-friendly synthesis routes. Integrating natural fibers and biodegradable substrates further promotes sustainable pathways for conductive, multifunctional textile development [14, 64, 71].

4.4 Mechanisms of Responsiveness

Smart textiles respond to external stimuli through a variety of mechanisms:

- **Thermo-responsiveness:** Thermo-responsive textiles are engineered to alter their physical or chemical behavior in response to temperature changes. This mechanism is often achieved by incorporating polymers with a lower critical solution temperature (LCST), such as poly(*N*-isopropylacrylamide), which undergoes reversible phase transitions between hydrophilic and hydrophobic states. This enables controlled breathability, moisture transport, and thermal insulation in fabrics. Microencapsulated phase-change materials also contribute by absorbing or releasing latent heat during solid–liquid transitions. The dynamic adjustment to thermal stimuli makes thermo-responsive textiles highly valuable for sportswear, protective clothing, and healthcare applications, offering adaptive comfort and energy-efficient performance under varying environmental conditions (e.g., thermochromic dyes, PCM-loaded fibers) [60].
- **Photo-responsiveness:** Photo-responsive textiles are functional fabrics that change their properties upon exposure to light, typically ultraviolet (UV) or visible radiation. This effect is often achieved by incorporating photochromic dyes, spiropyrans, or azobenzene-based molecules, which undergo reversible structural transformations when irradiated. Such molecular changes can alter fabric color, permeability, or surface wettability, enabling dynamic textile performance. Advanced applications include UV-protective garments, smart camouflage, and light-triggered drug delivery fabrics. Recent developments integrate nanostructures with photochromic agents to improve durability, wash resistance, and response speed, making photo-responsive textiles an emerging field in wearable technology, adaptive fashion, and protective clothing [1, 11, 73].

- **Moisture/Humidity responsiveness**: Moisture or humidity-responsive textiles are designed to adapt their structure or functionality in response to environmental water vapor. This behavior is typically achieved by incorporating hygroscopic polymers, natural fibers, or bilayer composites that swell, shrink, or bend with changes in relative humidity. Such textiles can regulate breathability, enhance wearer comfort, and act as passive actuators for smart garments. Recent advances employ cellulose nanofibers, hydrogels, and bio-inspired structures to develop fabrics with reversible dimensional changes and moisture-driven motion. Applications extend to self-ventilating clothing, sportswear, and wearable sensors, where humidity responsiveness provides adaptive performance without external energy input [37].
- **pH-responsiveness:** pH-responsive textiles are engineered by incorporating functional polymers, dyes, or nanomaterials that alter their physicochemical properties in response to environmental acidity or alkalinity changes. These materials typically rely on ionizable groups, such as carboxyl or amine functionalities, which undergo protonation or deprotonation, leading to swelling, color, or permeability variations. In textiles, pH-responsiveness enables smart applications, including wound dressings that indicate infection, drug-releasing fabrics, and reusable sensors for environmental monitoring. Recent strategies integrate natural biopolymers and nanostructures to enhance sensitivity, reversibility, and durability, promoting sustainable smart fabrics for biomedical, protective, and eco-friendly textile innovations [72].
- **Electro-responsiveness:** Electro-responsive textiles are designed to alter their properties or performance under an applied electric field. Integrating conductive polymers, metallic nanostructures, or carbon-based materials within textile substrates often achieves this functionality. When stimulated, these fabrics can undergo reversible changes in shape, conductivity, or permeability, enabling applications in soft actuators, artificial muscles, and wearable sensors. Electro-responsive behavior also supports energy storage devices, such as supercapacitor-integrated fabrics. Recent research emphasizes lightweight, flexible, and washable designs by combining electroactive polymers with sustainable fibers, paving the way for multifunctional smart textiles in healthcare, robotics, and adaptive clothing systems [7, 57].

The eco-friendly dimension of smart textiles emphasizes the integration of sustainability principles in material selection, processing, and end-of-life management. Traditional textile innovations often rely on synthetic polymers and energy-intensive treatments, raising concerns about environmental impact. Researchers are developing bio-based fibers, non-toxic functional finishes, and biodegradable nanomaterials that minimize ecological footprints to address this. Green synthesis of nanoparticles, waterless dyeing, and low-energy finishing technologies are also gaining traction. Furthermore, designing textiles for recyclability and extended durability aligns with circular economy goals. By merging smart functionality with sustainability, eco-friendly textiles provide a balanced pathway toward innovation, performance, and environmental responsibility [10].

4.5 Fabrication Techniques for Smart and Eco-Friendly Textiles

Fabrication of smart and eco-friendly textiles involves the careful integration of functional materials with sustainable processing methods. Conventional approaches such as dip-coating, padding, and screen-printing are widely used for embedding conductive polymers, nanoparticles, or responsive dyes, but modern techniques are increasingly focused on reducing environmental impact. Electrospinning enables the production of nanofibrous mats with high surface area and tailored responsiveness, while plasma treatment offers solvent-free surface modification. Layer-by-layer assembly and inkjet printing provide precise, low-waste functionalization for sensors and energy devices. Biopolymer composites and water-based processing methods are also being explored to minimize chemical residues. Combining smart functionality with green fabrication pathways ensures that future textiles meet technological and sustainability demands [15, 24].

4.5.1 Coating and Finishing

Coating and finishing are critical for imparting smart functionalities to textiles while ensuring durability and eco-compatibility. Conventional finishing often uses chemical treatments that enhance water repellency, flame resistance, or antimicrobial activity. However, emerging eco-friendly methods emphasize using bio-based coatings, non-toxic binders, and green solvents. Sol-gel processing, plasma-assisted finishing, and layer-by-layer deposition allow precise surface functionalization with minimal chemical waste. Functional nanoparticles, conductive polymers, and responsive dyes can be anchored onto fabrics to create self-cleaning, conductive, or protective textiles. Sustainable finishing also considers recyclability and wash durability, making coating technologies central to advancing smart, multifunctional, and environmentally responsible textile innovations [20]. Eco-friendly solutions involve water-based dispersions, sol-gel methods, or supercritical CO_2-assisted deposition to reduce chemical waste [54]. For example, ZnO nanoparticles deposited on cotton *via* a sol-gel route impart UV protection and antimicrobial activity, while water-based poly(3,4-ethylenedioxythiophene) (PEDOT): polystyrene sulfonate (PSS) coatings provide electrical conductivity without toxic solvents [61].

4.5.2 Electrospinning

Electrospinning is a versatile fabrication technique widely employed in developing smart and eco-friendly textiles. It uses a high-voltage electric field to draw ultrafine fibers from polymer solutions or melts, producing nonwoven mats with nanoscale diameters and high surface-area-to-volume ratios. This structure enables excellent breathability, flexibility, and responsiveness, making electrospun fabrics ideal for wearable sensors, filtration, drug delivery, and protective clothing applications. Incorporation of conductive polymers, nanoparticles, or bio-based materials further enhances their functional and sustainable performance. Electrospinning supports water-based and solvent-reduced processing, aligning with eco-friendly manufacturing goals. Recent advances focus on scalable electrospinning systems, hybrid nanofiber composites, and biodegradable polymers, ensuring that electrospun textiles combine high-tech functionality with environmental responsibility [1]. A key sustainability challenge is replacing petroleum-derived polymers with biodegradable alternatives. Recent studies demonstrate successful electrospinning of lignin-derived fibers with responsive coatings, paving the way for circular design [31, 69].

4.5.3 Weaving and Knitting with Conductive Yarns

Weaving and knitting with conductive yarns represent fundamental strategies for producing smart textiles that seamlessly integrate electronic functionality with comfort and flexibility. Conductive yarns, prepared by incorporating metallic fibers, conductive polymers, or carbon-based materials, can be interlaced into fabrics using conventional textile techniques without compromising structural integrity. Woven fabrics provide dimensional stability and durability, making them suitable for sensors, energy storage devices, and electromagnetic shielding. Knitted structures, on the other hand, offer stretchability and softness, enabling applications in wearable health monitors, motion detection systems, and interactive garments. The eco-friendly dimension is reinforced by using recycled fibers, biodegradable conductive coatings, and low-energy fabrication techniques. Current research emphasizes durability under repeated washing, scalability, and compatibility with sustainable fibers, ensuring conductive yarn integration supports both performance and environmental responsibility [51]. Eco-friendly innovations include silver-free conductive yarns (to avoid metal leaching) and coated with biopolymer-stabilized graphene [34].

4.5.4 Printing and Patterning

Printing and patterning provide efficient routes to functionalize textiles with smart properties while maintaining sustainability. Techniques such as screen printing, inkjet printing, and gravure printing allow precise deposition of conductive inks, responsive dyes, or nanomaterials, enabling applications in sensors, energy storage, and wearable electronics. Inkjet printing is beautiful due to its digital control, reduced material waste, and compatibility with water-based, eco-friendly inks. By integrating functional materials through low-waste, scalable methods, printing and patterning support the development of eco-friendly textiles that balance innovation, functionality, and environmental responsibility

[26]. Green innovation focuses on water-based inks, bio-derived colorants, and biodegradable binders to reduce ecological impact. Enzyme-assisted printing with natural dyes is gaining prominence [21, 50].

4.5.5 Layer-by-Layer Assembly

Layer-by-Layer (LbL) assembly is a versatile technique for functionalizing textiles with nanoscale precision while maintaining eco-friendly processing. It relies on the alternate deposition of oppositely charged polyelectrolytes, nanoparticles, or biomolecules, creating thin, uniform coatings with tunable thickness. This method enables the incorporation of flame-retardant, antimicrobial, conductive, or stimuli-responsive functionalities without altering the fabric's bulk properties. Since LbL often uses aqueous solutions and mild processing conditions, it aligns with green manufacturing principles. Recent advances integrate biopolymers, natural dyes, and nanocellulose in LbL coatings, ensuring sustainable smart textiles that combine durability, multifunctionality, and environmental compatibility for advanced wearable and protective applications [45].

4.5.6 3D and 4D Printing

3D and 4D printing technologies are emerging as transformative approaches in developing smart and sustainable textiles. 3D printing enables precise deposition of polymers, conductive inks, and bio-based composites onto fabrics, producing customizable structures with minimal waste. 4D printing extends this concept by incorporating stimuli-responsive materials that adapt their shape, color, or properties when exposed to heat, moisture, light, or electrical signals. These methods support the integration of sensors, actuators, and adaptive features directly into textiles. By using biodegradable filaments, water-based inks, and energy-efficient processing, 3D and 4D printing align with eco-friendly goals while expanding textile innovation and functionality [40]. For instance, eco-friendly approaches involve using PLA filaments, bio-inks derived from cellulose, and low-energy printing methods [46, 49, 56].

4.6 Green Chemistry in Smart Textile Development

Green chemistry principles are increasingly applied in smart textile development to minimize environmental impact while ensuring advanced performance. As discussed earlier, eco-friendly approaches include water-based or solvent-free processing, natural dyes, biodegradable polymers, and renewable nanomaterials such as cellulose, chitosan, or lignin. Catalysis and low-energy reactions enable functionalization of fabrics with antimicrobial, self-cleaning, or conductive properties under mild conditions. Greener surface modification techniques reduce hazardous chemicals and effluents, including plasma treatment and enzymatic finishing. Integrating smart functionality through sustainable routes improves recyclability and biodegradability and advances the vision of high-performance, eco-responsible textiles for wearable and protective applications [4].

4.6.1 Renewable and Bio-based Materials

To reduce the carbon footprint, the petroleum-based polymers are substituted with natural polymers such as cellulose, chitosan, silk, and alginate. They offer biodegradability, biocompatibility, and tunable functionality for sensors, actuators, and energy storage components. Lignin, starch, and protein-based derivatives are increasingly used as sustainable binders or conductive composites. Incorporating nanocellulose or bio-based nanoparticles enhances mechanical strength and responsiveness without compromising recyclability. Smart textiles can achieve multifunctionality while aligning with circular economy principles and reducing overall environmental footprints by integrating renewable feedstocks with advanced fabrication methods like electrospinning, coating, and 3D printing [6].

4.6.2 Non-toxic Functionalization

Non-toxic functionalization is a crucial strategy for developing safe smart textiles for both users and the environment. Traditional chemical treatments often employ heavy metals or toxic reagents such as formaldehyde-based resins, azo dyes, and halogenated flame retardants, which raise concerns regarding health hazards and ecological persistence. Eco-friendly alternatives rely on biopolymers, plant-based extracts, enzymatic processes, and water-based formulations to impart properties such as antimicrobial activity, UV protection, or conductivity. Plasma treatment and green nanoparticle synthesis further enable functionalization without harmful solvents. These approaches enhance durability and wash resistance while ensuring biodegradability and recyclability, positioning non-toxic functionalization as a cornerstone for sustainable innovation in smart and eco-friendly textiles [2].

4.6.3 Water- and Energy-Saving Processes

As discussed in the earlier chapters, water and energy-saving processes are central to sustainable textile manufacturing, as conventional dyeing and finishing consume vast resources and generate high pollution loads. Innovations such as supercritical CO_2 dyeing, digital printing, foam finishing, and plasma treatment significantly reduce water demand and eliminate harmful effluents. Low-liquor dyeing machines, enzymatic treatments, and microwave or infrared drying enhance energy efficiency. Integrating renewable energy sources into production also minimizes carbon footprints. Textiles can achieve high functionality by combining process intensification with cleaner technologies while conserving vital resources, aligning industrial practices with environmental protection and sustainable development goals.

4.6.4 Circular Economy Approaches

Circular economy approaches promote recycling, upcycling, and extended product lifecycles. Mechanical and chemical recycling of fibers, design for disassembly, and take-back schemes enable the recovery of valuable raw materials. Smart textiles increasingly adopt biodegradable polymers, renewable nanomaterials, and modular designs that support reuse and repair. Digital tools such as blockchain and material tracing enhance transparency in circular supply chains. By integrating eco-design principles with smart functionalities, circular economy approaches reduce environmental burdens while driving innovation toward a sustainable textile industry [8]. Figure 4.2 illustrates the circular economy in textile industries and smart textiles.

4.7 Applications of Smart and Responsive Eco-Friendly Textiles

Smart and responsive eco-friendly textiles are finding diverse applications across healthcare, sports, defense, and daily wear.

4.7.1 Healthcare and Biomedical Textiles

Biodegradable and biocompatible fibers such as chitosan, alginate, and cellulose nanofibers are widely used in wound dressings, drug delivery patches, and tissue engineering scaffolds. Smart textiles integrated with sensors enable continuous monitoring of vital signs, glucose levels, and wound healing progress. Antimicrobial and self-cleaning coatings derived from natural agents reduce hospital-acquired infections while ensuring patient safety. Eco-friendly biomedical fabrics minimize medical waste and environmental hazards, offering a sustainable alternative to conventional synthetic products in clinical care and personalized medicine [5, 10, 68].

Fig. 4.2 Circular economy in textile industries and smart textiles

4.7.2 Sports and Fitness Wear

Conductive yarns, flexible sensors, and energy-harvesting fabrics are integrated into garments to track parameters such as heart rate, respiration, hydration, and muscle activity in real time. Moisture-wicking and temperature-regulating eco-fibers improve comfort while reducing reliance on synthetic, non-biodegradable materials. Biodegradable polymers and recycled fibers lower environmental impact without compromising durability or elasticity. In addition, responsive coatings and nanomaterials allow clothing to adapt to sweat and temperature fluctuations, providing athletes with optimized performance, injury prevention, and improved recovery through sustainable wearable technologies [68].

4.7.3 Military and Protective Clothing

Smart eco-friendly textiles are increasingly adopted in military and protective clothing with flame-retardant, ballistic-resistant, and chemical-protective properties using biodegradable or recyclable polymers to reduce environmental impact. Responsive textiles can adapt to temperature, humidity, or light conditions, offering camouflage and thermal regulation in combat zones. Integration of sensors enables real-time monitoring of soldier health and environmental threats. Nanocoatings and bio-based finishes enhance durability, antimicrobial activity, and lightweight protection. By merging functionality with eco-conscious design, these textiles enhance operational performance while aligning with sustainability goals in defense applications [9].

4.7.4 Fashion and Consumer Wear

Smart eco-friendly textiles merge sustainability with personalization and functionality in fashion and consumer wear. Biodegradable fibers, recycled yarns, and plant-based dyes reduce environmental footprints while maintaining aesthetic appeal. Responsive fabrics enable color-changing effects, self-cleaning properties, and thermal comfort, enhancing consumer experience. Wearable electronics in garments allow connectivity, health tracking, and interactive lighting for interactive fashion. 3D and 4D printing further support customizable, waste-minimized production. By integrating circular economy principles with smart technologies, these textiles meet consumer demand for style and innovation and address pressing concerns of resource conservation and ethical manufacturing in the fashion industry [24].

4.7.5 Environmental Applications

Functionalized fabrics can filter microplastics, heavy metals, and pathogens from wastewater, while bio-based membranes are used in air filtration and capture of particulate matter. Textiles embedded with photocatalytic or conductive nanomaterials support self-cleaning surfaces and pollutant degradation under sunlight. Moisture-responsive fibers assist in water harvesting from the atmosphere, addressing freshwater scarcity. Additionally, biodegradable geotextiles stabilize soil and prevent erosion without leaving harmful residues. These innovations highlight the dual role of smart textiles in advancing environmental sustainability and addressing global ecological challenges [24].

4.8 Life Cycle Assessment (LCA) and Environmental Impact

Life Cycle Assessment (LCA) offers a structured approach to quantify the environmental impacts of smart and eco-friendly textiles throughout their entire lifecycle, from raw material production to end-of-life management. By examining factors such as energy consumption, water use, chemical inputs, and disposal pathways, LCA highlights critical areas where interventions can minimize environmental burdens. Incorporating renewable fibers, eco-friendly finishing agents, and recycling practices substantially reduces greenhouse gas emissions and waste compared to conventional textiles. Additionally, LCA facilitates the comparison of biodegradable, recyclable, and functionalized fabrics, ensuring that enhanced textile performance aligns with broader sustainability goals [59].

4.8.1 Raw Material Sourcing

Raw material sourcing is crucial in ensuring the sustainability of smart and eco-friendly textiles. Renewable materials such as organic cotton, hemp, bamboo, and regenerated cellulose fibers are increasingly preferred, reducing reliance on fossil-fuel-based synthetics. Recycling post-consumer products, including PET bottles and discarded garments, further improves resource efficiency and supports circular economy initiatives. Sustainable sourcing also prioritizes low-pesticide cultivation, water conservation, and ethical labor practices. Incorporating functional nanomaterials and biopolymers enhances the performance of textiles while preserving environmental integrity, thereby harmonizing technological innovation with ecological responsibility in textile production [19].

4.8.2 Manufacturing Phase

The manufacturing stage is a key contributor to the overall environmental impact of textiles. Fiber preparation, dyeing, finishing, and functionalization consume water, energy, and chemicals. Sustainable alternatives, such as digital printing, enzymatic treatments, plasma processing, and low-liquor dyeing, help minimize resource use and emissions. Careful assessment of energy sources, chemical hazards, and waste management is vital for accurate LCA analysis. By incorporating greener technologies and renewable inputs, the manufacturing phase can shift from an environmental burden to a driver of sustainability. For instance, plasma and supercritical CO_2 treatments can reduce water consumption by up to 80% compared to conventional wet processing [19].

4.8.3 Consumer Phase

The consumer phase typically represents the most considerable portion of a textile's environmental impact, mainly due to energy- and water-intensive activities such as washing, drying, and ironing. Conventional fabrics additionally release microfibers during laundering, contributing to water pollution. Smart and eco-friendly textiles mitigate these effects through antimicrobial, stain-resistant, and self-cleaning finishes, which reduce washing frequency and extend product lifespan. Thermo-regulating materials further decrease household energy consumption. Accurate LCA assessment requires consideration of user behavior, fabric durability, and maintenance practices, as these factors strongly influence the overall sustainability of textile products [29].

4.8.4 End-of-Life

Traditional disposal methods, such as landfilling and incineration, generate greenhouse gas emissions, contaminate soil, and result in the loss of valuable resources. Sustainable alternatives, including mechanical recycling, chemical fiber recovery, and biodegradation of natural or bio-based materials, significantly reduce these environmental impacts. Furthermore, innovative "design for disassembly" approaches enable separating embedded smart components from textiles, allowing materials to be reused or recycled effectively. These strategies help close the material loop, supporting circular economy principles and promoting the long-term sustainability of smart and eco-friendly textile products [59].

4.9 Digitalization and Industry 4.0 Integration

Smart textiles align closely with digitalization trends such as the Internet of Things (IoT), cloud computing, and AI-driven analytics. Integrating these aspects enhances their functionality while enabling sustainable management.

4.9.1 IoT-Enabled Smart Textiles

IoT-Enabled Smart Textiles can track physiological parameters such as heart rate, temperature, glucose level, and hydration levels. Further, transmitting this real-time data to smartphones or cloud platforms for healthcare, fitness, or safety applications helps individuals to get fit or track levels of disorders or diseases. IoT fabrics improve situational awareness and worker safety in industrial and military contexts. When combined with low-power electronics and energy-harvesting technologies, IoT textiles operate sustainably while minimizing battery dependency. Their role in eco-friendly innovation is enhancing product longevity, reducing resource-intensive monitoring equipment, and promoting a connected, responsive textile ecosystem [27, 41].

4.9.2 AI and Big Data

AI-driven models optimize fabric properties like thermal regulation or moisture management while reducing trial-and-error experimentation and resource use. Big Data collected from IoT-enabled textiles provides insights into user behavior, wear patterns, and environmental exposure, supporting adaptive textile functionalities. In manufacturing, AI enhances quality control and minimizes defects, lowering material waste. AI and Big Data foster closed-loop innovation, integrating sustainability with intelligence to drive the next generation of responsive, eco-conscious textile solutions [12].

4.9.3 Digital Twins for Textile Manufacturing

Digital twin technology revolutionizes textile manufacturing by creating real-time virtual replicas of machinery, processes, and production systems. These digital models continuously receive data from sensors embedded in equipment, enabling predictive maintenance, process optimization, and defect detection. In eco-friendly textile production, digital twins help simulate resource consumption, such as water and energy use, to minimize waste and emissions before physical trials. They also support adaptive design, allowing customization with reduced material loss. By integrating with AI and IoT platforms, digital twins enhance transparency, efficiency, and sustainability, making them central to Industry 4.0-driven textile innovations [3].

4.10 Durability, Washability, and Safety Issues

4.10.1 Washability and Wear Resistance

Washability and wear resistance are critical for the long-term usability of smart and eco-friendly textiles. Conventional laundering exposes fabrics to mechanical stress, detergents, and water, damaging conductive coatings, sensors, and functional finishes. Similarly, continuous wear leads to abrasion, bending, and environmental exposure, reducing performance. Researchers employ protective encapsulation, cross-linking agents, and nanocomposite coatings to address these challenges to preserve functionality during repeated washing cycles [53]. To counter this, cross-linking biopolymers and nanoencapsulation techniques improve wash durability without toxic binders [52].

4.10.2 Mechanical Robustness

Mechanical robustness is essential for maintaining the structural integrity of smart and eco-friendly textiles' and their functional performance under repeated stress. Such fabrics are subjected to stretching, bending, abrasion, and impact during daily use, which can degrade sensors, conductive yarns, and functional coatings. Enhancing mechanical robustness includes incorporating high-strength fibers, flexible conductive composites, nanofiller reinforcement, and cross-linked polymer coatings. Weaving, knitting, or electrospinning techniques can also optimize fiber orientation and distribution to improve load-bearing capacity. Ensuring mechanical durability extends the lifespan of smart textiles and supports sustainable practices by reducing premature disposal and resource waste [35, 36].

4.10.3 Toxicity and Biocompatibility

Toxicity and biocompatibility of wearable and biomedical smart and eco-friendly textiles are essential considerations in their application. If not chosen carefully, functional coatings, nanoparticles, and conductive materials may cause risks like skin irritation, cytotoxicity, and environmental accumulation. Biocompatible alternatives, such as natural polymers (cellulose, chitosan, silk), plant-based dyes, and green-synthesized nanoparticles, reduce health risks while maintaining effectiveness. Rigorous testing ensures safety, and prioritizing non-toxic, biodegradable materials supports sustainability and compliance, facilitating safe use in consumer and healthcare applications [16, 39, 43].

4.11 Challenges and Limitations

Despite significant advances, eco-friendly smart textiles face challenges, such as balancing durability, washability, and mechanical robustness, while integrating sensors, conductive yarns, and functional coatings is a significant challenge. Achieving high performance while ensuring environmental sustainability is complex, as some materials require energy-intensive production processes or contain non-biodegradable components. Additionally, cost, scalability, and compatibility with traditional textile manufacturing hinder commercialization. Regulatory and safety concerns must also be addressed, including toxicity, biocompatibility, and end-of-life management. To overcome these challenges, interdisciplinary research in materials science, green chemistry, and manufacturing technologies is essential for developing smart textiles that are multifunctional, safe, and sustainable [5, 32].

4.12 Future Outlook

The future of smart, eco-friendly textiles involves advanced materials, sustainable processing, and digital technologies. Innovations like biodegradable nanomaterials and conductive biopolymers will improve functionality while minimizing environmental impact. Integrating the Internet of Things (IoT), artificial intelligence (AI), and Industry 4.0 will enable adaptive and predictive textile applications. Adopting circular economy principles and green fabrication methods will enhance recyclability and reduce waste. Collaboration across materials science and textile engineering will drive the creation of safe, durable smart textiles, promoting a sustainable future for the industry. Overall, smart and responsive eco-friendly textiles represent a transformative frontier, merging sustainability, technology, and functionality into the fabrics of tomorrow.

4.13 Conclusion

Smart and eco-friendly textiles mark a transformative step in combining advanced functionality with sustainable practices and technological innovation. Innovations in conductive polymers, nanoparticles, biopolymers, and responsive fibers have expanded their use in healthcare, sports, defense, and environmental applications. These textiles gain enhanced adaptability and operational efficiency when integrated with modern technologies like IoT, artificial intelligence, and Industry 4.0 frameworks. Adopting green chemistry, renewable resources, and circular economy principles further reduces environmental impacts. While challenges in durability, washability, and large-scale production remain, ongoing research in materials science and manufacturing techniques promises safer, long-lasting, and environmentally responsible textile solutions, reflecting a shift toward high-performance, sustainable products that meet societal needs.

References

1. Aden, M. B., Rault, F., & Salaün, F. (2024). Photoresponsive Electrospun Polycaprolactone Nanofibrous Structure Embedding Dithienylethene Molecules. *Coatings*, *14*(6), 772. https://doi.org/10.3390/coatings14060772
2. Aguda, O. N., & Lateef, A. (2022). Recent advances in functionalization of nanotextiles: A strategy to combat harmful microorganisms and emerging pathogens in the 21st century. *Heliyon*, *8*(6), e09761. https://doi.org/10.1016/j.heliyon.2022.e09761
3. Alam, M. D., Kabir, G., & Mirmohammadsadeghi, S. (2023). A digital twin framework development for apparel manufacturing industry. *Decision Analytics Journal*, *7*, 100252. https://doi.org/10.1016/j.dajour.2023.100252
4. Arif, R., Jadoun, S., & Verma, A. (2021). Green chemistry in textile industry and their positive impact of implementation. In Green Chemistry for Sustainable Textiles (pp. 113–119). Elsevier. https://doi.org/10.1016/B978-0-323-85204-3.00023-3
5. Cherenack, K., & van Pieterson, L. (2012). Smart textiles: Challenges and opportunities. *Journal of Applied Physics*, *112*(9). https://doi.org/10.1063/1.4742728
6. Daget, T. M., kassie, B. B., & Tassew, D. F. (2025). A shift from synthetic to bio-based polymer for functionalization of textile materials: A review. *International Journal of Biological Macromolecules*, *306*, 141637. https://doi.org/10.1016/j.ijbiomac.2025.141637
7. Dai, H., Li, J., Jia, C., Zhai, Y., Tian, G., Wang, X., Xiang, H., & Zhu, M. (2025). Electro-Responsive Thermochromic and Mechanically Enhanced CNT Yarns through Solution Blow

Spinning Encapsulation. *ACS Applied Materials & Interfaces, 17*(17), 25951–25960. https://doi.org/10.1021/acsami.4c23025

8. De Felice, F., Fareed, A. G., Zahid, A., Nenni, M. E., & Petrillo, A. (2025). Circular economy practices in the textile industry for sustainable future: A systematic literature review. *Journal of Cleaner Production, 486*, 144547. https://doi.org/10.1016/j.jclepro.2024.144547
9. Degenstein, L. M., Sameoto, D., Hogan, J. D., Asad, A., & Dolez, P. I. (2021). Smart Textiles for Visible and IR Camouflage Application: State-of-the-Art and Microfabrication Path Forward *Micromachines, 12*(7). https://doi.org/10.3390/mi12070773
10. Dejene, B. K. (2025). The future of fabric: A comprehensive review of self-powered smart textiles and their emerging applications. *Energy Reports, 14*, 898–943. https://doi.org/10.1016/j.egyr.2025.07.002
11. Deng, C., Xue, J., Wang, J., Ren, D., Liu, G., Zhai, S., & Qi, D. (2025). Fabrication of photoresponsive self-deicing surface with micro-nano rough structures on fabrics. *Journal of Colloid and Interface Science, 683*, 845–857. https://doi.org/10.1016/j.jcis.2024.12.209
12. Dias, I. T. P. C., Borges, I. D. de S., & Valente, E. L. (2025). Artificial intelligence and the future of smart fabrics. *CONTRIBUCIONES A LAS CIENCIAS SOCIALES, 18*(2), e15665. https://doi.org/10.55905/revconv.18n.2-302
13. Dulal, M., Afroj, S., Ahn, J., Cho, Y., Carr, C., Kim, I.-D., & Karim, N. (2022). Toward Sustainable Wearable Electronic Textiles. *ACS Nano, 16*(12), 19755–19788. https://doi.org/10.1021/acsnano.2c07723
14. Ebadi, S. V., & Jafari, S. (2025). Tunable electrical conductivity and enhanced strain sensing performance of MXene/Polypyrrole textile sensors for wearable applications. *Surfaces and Interfaces, 60*, 106008. https://doi.org/10.1016/j.surfin.2025.106008
15. Ejegu, H., Jiang, X., Babeker, T. M. A., Kumah, C., Xu, M., Gebeyehu, E. K., Semanie, D. M., & Xu, G. (2025). Fabrication techniques and design considerations for fabric-based wearable soft systems and actuators. *Sensors and Actuators A: Physical, 395*, 117013. https://doi.org/10.1016/j.sna.2025.117013
16. Fanizza, C., Stefanelli, M., Risuglia, A., Bruni, E., Ietto, F., Incoronato, F., Marra, F., Preziosi, A., Mancini, P., Sarto, M. S., & Uccelletti, D. (2022). In Vitro and In Vivo Biocompatibility Studies on Engineered Fabric with Graphene Nanoplatelets. *Nanomaterials (Basel, Switzerland), 12*(9). https://doi.org/10.3390/nano12091405
17. Farooq, A. S., & Zhang, P. (2021). Fundamentals, materials and strategies for personal thermal management by next-generation textiles. *Composites Part A: Applied Science and Manufacturing, 142*, 106249. https://doi.org/10.1016/j.compositesa.2020.106249
18. Farouk, A., Sharaf, S., & Abd El-Hady, M. M. (2013). Preparation of multifunctional cationized cotton fabric based on TiO2 nanomaterials. *International Journal of Biological Macromolecules, 61*, 230–237. https://doi.org/10.1016/j.ijbiomac.2013.06.022
19. Fonseca, A., Ramalho, E., Gouveia, A., Henriques, R., Figueiredo, F., & Nunes, J. (2023). Systematic Insights into a Textile Industry: Reviewing Life Cycle Assessment and Eco-Design. *Sustainability, 15*(21), 15267. https://doi.org/10.3390/su152115267
20. Ghosh, J., Rupanty, N. S., Noor, T., Asif, T. R., Islam, T., & Reukov, V. (2025). Functional coatings for textiles: advancements in flame resistance, antimicrobial defense, and self-cleaning performance. *RSC Advances, 15*(14), 10984–11022. https://doi.org/10.1039/D5RA01429H
21. Haddar, W., Adeel, S., Salman, M., Ghaffar, A., Naseer, M., Usama, M., & Ben Ticha, M. (2023). Ecological Applications of Enzymes in Plants Based Textile Dyeing. In Dyes and Pigments—Insights and Applications. IntechOpen. https://doi.org/10.5772/intechopen.108603
22. Haque, A. N. M. A., Zhang, Y., Cai, Z., Lee, E., & Naebe, M. (2024). Conductive washable wool textiles: Evaluating graphene oxide coating methods. *Applied Materials Today, 41*, 102476. https://doi.org/10.1016/j.apmt.2024.102476

23. Hardy, J. G., Lee, J. Y., & Schmidt, C. E. (2013). Biomimetic conducting polymer-based tissue scaffolds. *Current Opinion in Biotechnology, 24*(5), 847–854. https://doi.org/10.1016/j.copbio.2013.03.011
24. Hossain, M. T., Shahid, M. A., Limon, M. G. M., Hossain, I., & Mahmud, N. (2024). Techniques, applications, and challenges in textiles for a sustainable future. *Journal of Open Innovation: Technology, Market, and Complexity, 10*(1), 100230. https://doi.org/10.1016/j.joitmc.2024.100230
25. Hou, Z., Liu, X., Tian, M., Zhang, X., Qu, L., Fan, T., & Miao, J. (2023). Smart fibers and textiles for emerging clothe-based wearable electronics: materials, fabrications and applications. *Journal of Materials Chemistry A, 11*(33), 17336–17372. https://doi.org/10.1039/D3TA02617E
26. Htwe, Y. Z. N., & Mariatti, M. (2022). Printed graphene and hybrid conductive inks for flexible, stretchable, and wearable electronics: Progress, opportunities, and challenges. *Journal of Science: Advanced Materials and Devices, 7*(2), 100435. https://doi.org/10.1016/j.jsamd.2022.100435
27. Iman, U. R., Zada, M., Basir, A., Hayat, S., Lim, Y.-H., & Yoo, H. (2024). IoT-Enabled Real-Time Health Monitoring via Smart Textile Integration With LoRa Technology Across Diverse Environments. *IEEE Transactions on Industrial Informatics, 20*(11), 12803–12813. https://doi.org/10.1109/TII.2024.3424517
28. Islam, T., Hossain, M. M., & Covington, S. (2025). Natural Fibers Towards Fashion Sustainability: A Review of Raw Materials, Production, Application, and Perspective. *Journal of Natural Fibers, 22*(1). https://doi.org/10.1080/15440478.2025.2462218
29. Jacometti, V. (2019). Circular Economy and Waste in the Fashion Industry. *Laws, 8*(4), 27. https://doi.org/10.3390/laws8040027
30. Jafarpisheh, F., Kolvir, H. R., & Ghorbanpour, M. (2024). Photocatalytic self-cleaning and antibacterial activity of modified wood by Ag doped TiO2 nanoparticles. *Next Materials, 4*, 100207. https://doi.org/10.1016/j.nxmate.2024.100207
31. Jiju, P. S., Patel, A. K., Shruthy, N. S., Shalu, S., Dong, C.-D., & Singhania, R. R. (2025). Sustainability through lignin valorization: recent innovations and applications driving industrial transformation. *Bioresources and Bioprocessing, 12*(1), 88. https://doi.org/10.1186/s40643-025-00929-x
32. Júnior, H. L. O., Neves, R. M., Monticeli, F. M., & Dall Agnol, L. (2022). Smart Fabric Textiles: Recent Advances and Challenges. *Textiles, 2*(4), 582–605. https://doi.org/10.3390/textiles2040034
33. Karimah, A., Ridho, M. R., Munawar, S. S., Adi, D. S., Ismadi, Damayanti, R., Subiyanto, B., Fatriasari, W., & Fudholi, A. (2021). A review on natural fibers for development of eco-friendly bio-composite: characteristics, and utilizations. *Journal of Materials Research and Technology, 13*, 2442–2458. https://doi.org/10.1016/j.jmrt.2021.06.014
34. Keawploy, N., Venkatkarthick, R., Wangyao, P., Zhang, X., Liu, R., & Qin, J. (2020). Eco-Friendly Conductive Cotton-Based Textile Electrodes Using Silver- and Carbon-Coated Fabrics for Advanced Flexible Supercapacitors. *Energy & Fuels, 34*(7), 8977–8986. https://doi.org/10.1021/acs.energyfuels.0c01419
35. Kim, J., Choi, Y., Jang, H., Jiong, S., Chen, X., Seo, B., & Choi, W. (2024). Thermo-Chemo-Mechanically Robust, Multifunctional MXene/PVA/PAA-Hanji Textile with Energy Harvesting, EMI Shielding, Flame-Retardant, and Joule Heating Capabilities. *Advanced Materials, 36*(47). https://doi.org/10.1002/adma.202411248
36. Liu, Q., Zhang, Y., Sun, X., Liang, C., Han, Y., Wu, X., & Wang, Z. (2023). All textile-based robust pressure sensors for smart garments. *Chemical Engineering Journal, 454*, 140302. https://doi.org/10.1016/j.cej.2022.140302

37. Liu, R., Wang, Y., Fan, W., Zou, J., Yu, X., Liu, Y., & Zhi, C. (2024). Adaptive dynamic smart textiles for personal thermal-moisture management. *European Polymer Journal, 206*, 112777. https://doi.org/10.1016/j.eurpolymj.2024.112777
38. Liu, X., Xu, F., Yong, H., Chen, D., Tang, C., Kan, J., & Liu, J. (2025). Recent advances in chitosan-based active and intelligent packaging films incorporated with flavonoids. *Food Chemistry: X, 25*, 102200. https://doi.org/10.1016/j.fochx.2025.102200
39. Machado, V., Marinho, A., Vieira de Castro, P., & Silva, T. (2025). From Fabric to Finish: The Cytotoxic Impact of Textile Chemicals on Humans Health. *Textiles, 5*(2), 16. https://doi.org/10.3390/textiles5020016
40. Manaia, J. P., Cerejo, F., & Duarte, J. (2023). Revolutionising textile manufacturing: a comprehensive review on 3D and 4D printing technologies. *Fashion and Textiles, 10*(1), 20. https://doi.org/10.1186/s40691-023-00339-7
41. Manglani, H., Hodge, G. L., & Oxenham, W. (2019). Application of the Internet of Things in the textile industry. *Textile Progress, 51*(3), 225–297. https://doi.org/10.1080/00405167.2020.1763701
42. Mohandas, A., Deepthi, S., Biswas, R., & Jayakumar, R. (2018). Chitosan based metallic nanocomposite scaffolds as antimicrobial wound dressings. *Bioactive Materials, 3*(3), 267–277. https://doi.org/10.1016/j.bioactmat.2017.11.003
43. Mondal, M., Banerjee, S., Banerjee, S., Naik, S. N., & Ali, S. W. (2024). *Biocompatibility and Toxicity of Nanoparticles in Textile Finishing* (pp. 461–486). https://doi.org/10.1007/978-981-97-2696-7_16
44. Mou, L., Li, J., Lu, Y., Li, G., & Li, J. (2025). Polylactic acid: A future universal biobased polymer with multifunctional performance—from monomer synthesis, and processing to applications: A review. *Journal of Hazardous Materials Advances, 18*, 100757. https://doi.org/10.1016/j.hazadv.2025.100757
45. Pan, Y., Fu, L., Du, J., Zhang, D., Lü, T., Zhang, Y., & Zhao, H. (2022). Layer-by-Layer Self-Assembly Coating for Multi-Functionalized Fabrics: A Scientometric Analysis in CiteSpace (2005–2021). *Molecules, 27*(19), 6767. https://doi.org/10.3390/molecules27196767
46. Patel, P. K., Sonagara, C., Kumar, S., Meena, K., & Banerjee, S. S. (2025). 3D Printing of Bioplastics for Medical Applications. *Polymer Engineering & Science, 65*(8), 3880–3903. https://doi.org/10.1002/pen.27266
47. Pereira, P. H. F., Costa, S. A., Costa, S. M., & Arantes, V. (2025). Efficient production of lignin nanoparticle colloids and their feasibility for eco-friendly dyeing of natural and synthetic textile fabrics. *Industrial Crops and Products, 225*, 120390. https://doi.org/10.1016/j.indcrop.2024.120390
48. Perelshtein, I., Applerot, G., Perkas, N., Wehrschetz-Sigl, E., Hasmann, A., Guebitz, G. M., & Gedanken, A. (2009). Antibacterial Properties of an In Situ Generated and Simultaneously Deposited Nanocrystalline ZnO on Fabrics. *ACS Applied Materials & Interfaces, 1*(2), 361–366. https://doi.org/10.1021/am8000743
49. Pramanik, J., Brahma, B., Pradhan, S., Ranjan Senapati, M., Kumar Samal, A., & Kumar Pani, S. (2024). 3D printing application in biomedical—a review. *Materials Today: Proceedings, 115*, 95–100. https://doi.org/10.1016/j.matpr.2023.07.046
50. Rahaman, Md. T., Khan, Md. S. H., & Al Mamun, Md. A. (2025). Green Extraction of Natural Pigments and Eco-Friendly Binders for Textile Coloration: Sources, Applications, Challenges, and Sustainability Issues. *Materials Circular Economy, 7*(1), 29. https://doi.org/10.1007/s42824-025-00168-7
51. Repon, M. R., Laureckiene, G., & Mikucioniene, D. (2021). The Influence of Electro-Conductive Compression Knits Wearing Conditions on Heating Characteristics. *Materials (Basel, Switzerland), 14*(22). https://doi.org/10.3390/ma14226780

52. Rodtuk, N., Kamlangmak, N., Sajomsang, W., & Chaiyasat, A. (2025). Multifunctional Biopolymer Nanocapsules for Antimicrobial, Fragrance-Releasing, and UV-Activated Coatings. *ACS Omega, 10*(28), 30829–30842. https://doi.org/10.1021/acsomega.5c03118
53. Rotzler, S., & Schneider-Ramelow, M. (2021). Washability of E-Textiles: Failure Modes and Influences on Washing Reliability. *Textiles, 1*(1), 37–54. https://doi.org/10.3390/textiles1010004
54. Rungruangkitkrai, N., Phromphen, P., Chartvivatpornchai, N., Srisa, A., Laorenza, Y., Wongphan, P., & Harnkarnsujarit, N. (2024). Water Repellent Coating in Textile, Paper and Bioplastic Polymers: A Comprehensive Review. *Polymers, 16*(19), 2790. https://doi.org/10.3390/polym16192790
55. Shah, M. A., Pirzada, B. M., Price, G., Shibiru, A. L., & Qurashi, A. (2022). Applications of nanotechnology in smart textile industry: A critical review. *Journal of Advanced Research, 38*, 55–75. https://doi.org/10.1016/j.jare.2022.01.008
56. Sharma, C., Raza, M. A., Purohit, S. D., Pathak, P., Gautam, S., Corridon, P. R., & Han, S. S. (2025). Cellulose-based 3D printing bio-inks for biomedical applications: A review. *International Journal of Biological Macromolecules, 305*, 141174. https://doi.org/10.1016/j.ijbiomac.2025.141174
57. Sheng, M., Zhang, L., Jiang, S., Yang, L., Zaaboul, F., & Fu, S. (2021). Bioinspired Electro-Responsive Multispectral Controllable Dye-Doped Liquid Crystal Yolk–Shell Microcapsules for Advanced Textiles. *ACS Applied Materials & Interfaces, 13*(11), 13586–13595. https://doi.org/10.1021/acsami.1c00003
58. Stoppa, M., & Chiolerio, A. (2014). Wearable Electronics and Smart Textiles: A Critical Review. *Sensors, 14*(7), 11957–11992. https://doi.org/10.3390/s140711957
59. Subramanian Senthilkannan Muthu. (2015). *Handbook of Life Cycle Assessment (LCA) of Textiles and Clothing*. Elsevier. https://doi.org/10.1016/C2014-0-00761-7
60. Tang, K. H. D. (2025). Advances in Thermoregulating Textiles: Materials, Mechanisms, and Applications. *Textiles, 5*(2), 22. https://doi.org/10.3390/textiles5020022
61. Tania, I. S., & Ali, M. (2021). Coating of ZnO Nanoparticle on Cotton Fabric to Create a Functional Textile with Enhanced Mechanical Properties. *Polymers, 13*(16), 2701. https://doi.org/10.3390/polym13162701
62. Tao, X. (2001). Smart technology for textiles and clothing—introduction and overview. In Smart Fibres, Fabrics and Clothing (pp. 1–6). Elsevier. https://doi.org/10.1533/9781855737600.1
63. Tian, Y., Liu, Y., Zhang, L., Hua, Q., Liu, L., Wang, B., & Tang, J. (2020). Preparation and characterization of gelatin-sodium alginate/paraffin phase change microcapsules. *Colloids and Surfaces A: Physicochemical and Engineering Aspects, 586*, 124216. https://doi.org/10.1016/j.colsurfa.2019.124216
64. Tseghai, G. B., Mengistie, D. A., Malengier, B., Fante, K. A., & Van Langenhove, L. (2020). PEDOT:PSS-Based Conductive Textiles and Their Applications. *Sensors, 20*(7), 1881. https://doi.org/10.3390/s20071881
65. Vasantharaj, S., Sathiyavimal, S., Saravanan, M., Senthilkumar, P., Gnanasekaran, K., Shanmugavel, M., Manikandan, E., & Pugazhendhi, A. (2019). Synthesis of ecofriendly copper oxide nanoparticles for fabrication over textile fabrics: Characterization of antibacterial activity and dye degradation potential. *Journal of Photochemistry and Photobiology B: Biology, 191*, 143–149. https://doi.org/10.1016/j.jphotobiol.2018.12.026
66. Vigneshwaran, N., Kathe, A. A., Varadarajan, P. V., Nachane, R. P., & Balasubramanya, R. H. (2007). Silver−Protein (Core−Shell) Nanoparticle Production Using Spent Mushroom Substrate. *Langmuir, 23*(13), 7113–7117. https://doi.org/10.1021/la063627p

67. Wang, H., Zhao, L., Song, G., Tang, G., & Shi, X. (2018). Organic-inorganic hybrid shell microencapsulated phase change materials prepared from SiO2/TiC-stabilized pickering emulsion polymerization. *Solar Energy Materials and Solar Cells, 175*, 102–110. https://doi.org/10.1016/j.solmat.2017.09.015
68. Xu, Z., Zhang, C., Wang, F., Yu, J., Yang, G., Surmenev, R. A., Li, Z., & Ding, B. (2025). Smart Textiles for Personalized Sports and Healthcare. *Nano-Micro Letters, 17*(1), 232. https://doi.org/10.1007/s40820-025-01749-6
69. Yadav, R., Zabihi, O., Fakhrhoseini, S., Nazarloo, H. A., Kiziltas, A., Blanchard, P., & Naebe, M. (2023). Lignin derived carbon fiber and nanofiber: Manufacturing and applications. *Composites Part B: Engineering, 255*, 110613. https://doi.org/10.1016/j.compositesb.2023.110613
70. Yang, M., Pan, J., Xu, A., Luo, L., Cheng, D., Cai, G., Wang, J., Tang, B., & Wang, X. (2018). Conductive Cotton Fabrics for Motion Sensing and Heating Applications. *Polymers, 10*(6), 568. https://doi.org/10.3390/polym10060568
71. Zahid, M., Anwer Rathore, H., Tayyab, H., Ahmad Rehan, Z., Abdul Rashid, I., Lodhi, M., Zubair, U., & Shahid, I. (2022). Recent developments in textile based polymeric smart sensor for human health monitoring: A review. *Arabian Journal of Chemistry, 15*(1), 103480. https://doi.org/10.1016/j.arabjc.2021.103480
72. Zhang, Yangyang, & Gong, X. (2023). Smart and durable pH-responsive superhydrophobic fabrics with switchable surface wettability for high-efficiency and complex oil/water separation. *Giant, 14*, 100157. https://doi.org/10.1016/j.giant.2023.100157
73. Zhang, Ye, Wang, X., Zhang, X., Wang, W., Yao, Y., Pan, J., Shao, G., Bi, S., Chen, N., Jiang, J., & Shao, H. (2025). Tunable photo-responsive liquid crystal elastomer fibers disperse dyeing for smart textiles. *Materials Horizons, 12*(13), 4862–4874. https://doi.org/10.1039/D5MH00357A

MIX
Papier aus verantwortungsvollen Quellen
Paper from responsible sources
FSC® C105338

If you have any concerns about our products,
you can contact us on
ProductSafety@springernature.com

In case Publisher is established outside the EU,
the EU authorized representative is:
Springer Nature Customer Service Center GmbH
Europaplatz 3, 69115 Heidelberg, Germany

Printed by Libri Plureos GmbH
in Hamburg, Germany